LIGHT

LIGHT

Our Bridge to the Stars

JOHN RUBLOWSKY

Foreword by WILLY LEY
Illustrated by Nancy Grossman

BASIC BOOKS, INC.
PUBLISHERS
New York   London

© 1964 by John Rublowsky
Library of Congress Catalog Number: 64-17280
Manufactured in the United States of America
Designed by Loretta Li

To Stefan

FOREWORD

Willy Ley

This book might also have the title, "A Case History of Light," because here we have the thoughts evolved through the centuries around the apparently simple phenomenon of light.

At first, light was just light. It was taken for granted during the day and artificially created at night to imitate the day within a small space. It thus set the real and imaginary fears of primitive man to rest. But soon questions began to occur to people at a more advanced stage of thought. For example, if I light a lantern and it begins to shed light and if there is an observer a mile away, will the light of the lantern be visible to him as soon as it is to me, or will he see it a little later? In other words, does light need time to travel from one place to another? If so, how much?

Of course, light was used to carry messages from one place to another. But in the nineteenth century there came the surprising and wonderful discovery that light carries a message of its own, namely, the information of what substance it is that is glowing at a distance.

In Mr. Rublowsky's book, we have the story of light and of how mankind learned to understand it, even though our understanding is not yet complete. And this story is also the answer to a question so often asked by laymen, namely, why scientists say what they say about the stars. How can they know? After all, they just see or photograph a few pinpoints of light in the sky. By the time you reach the last chapter, you will know the answer to that question.

CONTENTS

FOREWORD	*Willy Ley*	vii
LIST OF ILLUSTRATIONS		x
1	"LET THERE BE LIGHT"	1
2	THE SOURCE OF LIGHT	13
3	THE SPECTRUM	26
4	PRACTICAL APPLICATIONS	39
5	THE WAY TO THE STARS	50
6	THE LANGUAGE OF LIGHT	60
7	BREAKING THE CODE	71
8	WINDOW BEYOND SIGHT	82
9	COHERENT LIGHT, THE NEW TOOL	92
10	IN WONDERLAND	102
11	THE UNIVERSAL CONSTANT	115
GLOSSARY		125
CAPSULE BIOGRAPHIES		133
INDEX		141

LIST OF ILLUSTRATIONS

CHAPTER		PAGE
1	Olaus Roemer (1644-1710)	7
2	Sun Symbols	15
3	René Descartes (1596-1650)	30
4	Galileo Galilei (1564-1642)	45
5	Isaac Newton (1642-1727)	56
6	Wilhelm Konrad Roentgen (1845-1923)	67
7	Joseph von Fraunhofer (1787-1826)	79
8	Karl G. Jansky (1905-1950)	85
9	Niels Bohr (1885-1962)	95
10	James Clerk Maxwell (1831-1879)	103
11	Albert Einstein (1879-1955)	120

LIGHT

1

"LET THERE BE LIGHT"

> ... And darkness was upon the face of the deep. ...
> And God said, Let there be light. ... And there was
> Light. ... And God saw the light, that it was good. ...

These words, written some 5,000 years ago by Hebrew shepherds trying to understand and explain the bewildering universe around them, reveal the importance of light in the life of man. The ancients considered light so vital to their lives that they believed that God brought it forth on the first day of creation, along with the heavens and the earth.

To this day, all the peoples of the world pay homage to light. Votive candles play an important part in almost every religious ritual, from the Buddhist ceremony to the Catholic mass. Many Jewish religious ceremonies begin with the lighting of a candle. A perpetual flame burns before the Tomb of the Unknown Soldier under the *Arc de Triomphe* in Paris.

We pay unconscious tribute to light even in our daily lives. This tribute is ingrained in our very language. When we do not know something, we say that "we are in the dark." We overcome our ignorance by "throwing *light* on the subject." Darkness is mysterious, unfathomable, and frightening. It is bound up intimately with evil and melancholy feelings. The *Prince of Darkness* is the devil. When we are feeling badly and out-of-sorts, we say that we are "in a *dark* mood." When we are really upset, our mood becomes *black*. There is *black magic*, which is evil. And many of us still shy away from the sight of a *black cat*.

Light, on the other hand, represents everything that is good, open, and cheerful. The very word sounds airy and weightless. We are *light-hearted* when things go well for us. We see *light in the dark* when we have solved a baffling problem.

It is not difficult to understand why light is so important in our lives. Light is the essence of the vision with which we perceive the world around us. It is everywhere; it determines everything we see. It is the most commonplace thing in our lives; at the same time, it is also one of the most mysterious. We cannot see without light, and yet we cannot see light as it travels through space. Though it is such an accepted part of our daily lives, light is actually a visitor that comes to us across a great, alien gulf of space. Nearly all the light we use has come to our eyes as invisible rays of energy traveling some 93,000,000 miles from the sun.

Light is the conveyer of form and color, and, along with the rest of the electromagnetic spectrum that ranges from the tiniest cosmic rays to mile-long radio waves, it is the swiftest carrier of energy in the universe. Its speed is the absolute limit of propagation of material influence anywhere, for Einstein's relativity equations show that nothing can move faster than the speed of light.

If you are reading by the light of day, the light you see reflected from this page was actually inside the sun 7.8 minutes ago. Each photon of this light spent something

"Let There Be Light"   3

like ten thousand years milling wildly about on its way to the solar surface before it began its swift journey earthward.

Light is at once material and immaterial. When it passes through a crystal, it is dispersed into a pattern that could be produced only by an immaterial wave. Yet, it is also so substantial that it will literally fall of its own weight if unsupported or bend under the stress of gravity from a star. It is heavy enough to turn the tail of a comet so that it is forever pointing away from the sun and the steady flow of light that pours from the sun's surface.

Light is so potent that a burst of it from a single flash bulb can kill a rat. A slightly longer exposure from an ordinary light bulb will stop certain plants from flowering or trigger the beginning of a bird's song. Light can be twisted like a cable or pumped through valves like a gas.

Light can even be described in terms of our most familiar economic units, as was suggested by the engineer who figured out that the light which is produced by man on earth costs about $400,000,000 a pound, delivered. This gives us some idea of the riches lavished on us by the sun. Its rays deliver just about one pound of light per square mile every day across the entire surface of the earth.

Light, as we shall see, is also closely tied up with the very essence of matter. It now offers physicists a tool with which to explore the innermost depths of the atom, even though it is so mysterious and baffling a phenomenon that it is only within the past few decades that scientists have really begun to understand the nature of light.

Scientists, however, need not fully understand a thing before they can use that thing. Electricity, heat, and gravity are all examples of natural phenomena that scientists put to good use long before they were fully, or even partially, understood. The same holds true for light. Its qualities and properties were studied and used long before any scientist could tell us what light really was.

How did the origin and properties of light come to be explored? Although the religious explanation that God had said, "Let there be light," and there was light was accepted for thousands of years, it did not satisfy many curious people. They wanted to know more. They wanted to understand exactly what light was, where and how it was produced, and what made it glow in so many beautiful colors.

Indeed, the phenomenon of light proved so intriguing to scientists that it attracted the attention of the greatest minds in the world. Aristotle, Bacon, Galileo, Grimaldi, Newton, Hooke, Huygens, Descartes, Young, Foucault, Olbers, Maxwell, Hertz, de Broglie, Rutherford, Einstein, Dirac, Fermi, Moseley, Bohr—these are just a few of the great scientists who have helped man unravel the mysteries of light.

As early as the fifth century B.C., men had already begun to think seriously about the nature of light. Empedocles, a philosopher then living in Greece, first proposed the modern-sounding theory that light, since it moves from one point to another, must travel at a measurable rate of speed. Empedocles' ideas could not, of course, have been based on scientific observation. He did not have the tools or the instruments with which to make an accurate study of light. His ideas, more likely, were based solely on reasoning and intuition, the source of most knowledge and theory in those early days of scientific development.

A hundred years later, another Greek philosopher, Aristotle, also considered the nature of light and added his own logical speculations to the store of scientific knowledge. He wondered about the nature of color and believed

—wrongly as we now know—that white light was pure. Colors, according to his theory, were merely the results of the pure, transcendental white light suffused with earthly properties.

Aristotle also wondered about the movement of light. He agreed with Empedocles' logic, but added his own speculations to his predecessor's theories. "If light does take time to move from one point to another," reasoned Aristotle, "then it follows that any given time is divisible into parts, so that we can assume a time when the sun's rays were not as yet seen but were still traveling in the middle space before it reaches the earth."

He further reasoned that light had to travel very swiftly, especially through the "middle space between heaven and earth." Once it reached the earth, however, its speed was slowed considerably, especially after the pure light of heaven became infected with earthly properties and changed colors.

Aristotle's theories, though they proved wrong in many ways, became a landmark in the journey toward the understanding of the nature of light, although, ironically enough, they could be neither confirmed nor disproved for nearly two thousand years. Light was simply too flimsy an entity to lend itself to easy study.

In 1667, Galileo Galilei tried to measure the speed of light in one of the first truly scientific attempts to study a natural phenomenon. Galileo was dissatisfied with Aristotle's pronouncement that light travels very swiftly. He wanted to find out exactly how swiftly it moved. In order to do so, Galileo conducted an ingenious experiment. He had two men with very bright lanterns flash a beam back and forth to each other from the tops of two high hills on opposite sides of a broad valley. Galileo posted himself on a third hill from which he could observe both lanterns. For weeks, the men practiced, each one uncovering his light as quickly as possible when he saw the lantern beam from the hill on the other side of the valley.

When Galileo was satisfied that both men were equally swift in returning the signal, he took up his station with a few other observers. One man uncovered his lantern, and light flashed to the second hill. There, the second lantern twinkled its answer. Though they repeated the experiment many times, neither Galileo nor the other observers with him could tell how fast the second light appeared. Light moved too quickly to be measured in this crude manner.

Soon after, however, a method was discovered whereby the speed of light was actually measured with a fair degree of accuracy. Eight years after Galileo flashed his lanterns in vain, the Danish astronomer Olaus Roemer produced the first conclusive evidence that both Empedocles and Aristotle were right when they said that light traveled at a measurable speed. In order to find this proof, Roemer had to go completely out of this world. This was not a difficult thing for him to do because, dedicated astronomer that he was, Roemer spent most of his waking hours staring at the stars through a telescope patterned after the instruments Galileo had made some ten years earlier.

Over four years, Roemer had been keeping precise records of the eclipses of Io, the innermost of the four known moons of Jupiter. He timed these eclipses carefully with a new kind of pendulum clock that he had designed himself. This clock was much more accurate than any other timepiece of his age, and with it Roemer could make very accurate observations.

Io, the astronomer discovered, routinely disappears behind giant Jupiter every 42 hours, 28 minutes, with only one puzzling discrepancy. Io had the peculiar habit of always being a little late in ducking behind Jupiter during each half-year that the earth pulled away from the big planet and then speeding up progressively during the next six months as the earth pulled toward Jupiter in the course of the earth's orbital journey.

OLAUS ROEMER (1644-1710)

At first, Roemer believed that this strange sequence resulted from a fault in his clock. He checked and rechecked the clock, but could find no fault. The difference in timing, he finally decided, was caused by the different distances the light from Io had to travel in order to reach earth during different times of the year. On the basis of this carefully gathered evidence, Roemer concluded that the light from Io must take about twenty-two minutes to travel the distance across the diameter of earth's orbit. This was the difference in the time of Io's eclipse as measured from opposite sides of the earth's orbital path.

With this data, it was not difficult for Roemer to compute the speed of light. According to his calculations, this speed worked out to be a staggering 140,000 miles per second. When one considers the conditions under which Roemer worked, his findings were remarkable. We know today that the speed of light is 186,200 miles per second. He erred by less than 15 per cent.

Roemer's error, however, was due mainly to the fact that he did not know the exact diameter of earth's orbit. He believed that it was somewhat smaller than it actually is. Because of this error, Roemer could not know how far the light reflected from Io really traveled before it reached his telescope.

The Danish astronomer's discovery was very exciting to the scientists of his day. Eager astronomers, professional and amateur, turned their telescopes on Jupiter to time the eclipses of the moon Io according to the new calculations of Roemer. The distant satellite moved closer and closer to Jupiter, and, at precisely the moment predicted by Roemer, it slipped from view behind the planet. In this way, the moon of Jupiter, first seen by Galileo Galilei, proved what the old Italian scientist could not—that the speed of light was finite, that it could be measured and timed.

"Let There Be Light"

After Roemer's demonstration of the measurability of the speed of light, scientists all over the world turned to the study of light with renewed interest and vigor. This profound interest in light and its remarkable properties continues to this day, as witnessed by the magnificent structure of modern astronomy. Astronomy, which deals most specifically with light, is the oldest branch of science. Indeed, it is as old as history itself, for men have always seen the brilliant panoply of night and stared at the stars in wonder. Light is at the very heart of astronomy.

Earth is separated from the stars and even the other planets by vast distances. There is but one link between us and the other heavenly bodies, and this link consists only of the tenuous waves of light that shine down upon us from the farthest reaches of space. Light is the bridge. All that we know about the stars and the planets we have learned through light.

Consider, for a moment, the universe as it is revealed by modern astronomy. It is so grand as to stagger the imagination. Our earth and even our sun, which both loom so large when measured in human terms, are reduced in this universal scale to mere grains of sand in the infinite reaches of empty space. Our proud sun, once worshiped as a god, is revealed for what it actually is—a minor star, one among a *hundred billion,* in one of the spiral arms of an ordinary galaxy. Stretching away from our galaxy, as far as our most powerful telescopes can see, are more galaxies and even huge systems of galaxies, each made up of billions on billions of stars. And these galaxies are rushing away from one another at speeds which approach that of light itself. Over the years, astronomers have discovered order in this seeming chaos of swirling stars and galaxies, but this order is so complex that it can be understood and expressed only in terms of the most abstract mathematics.

And, if the scope of this universe staggers the imagination, the details painstakingly ferreted out by astronomers are equally awesome. Over a thousand different classes of stars have been identified, and their variety in appearance, mass, and luminosity is startling. A star called S Doradus in the larger Magellanic Cloud, for example, is 400 billion times as bright as the dimmest star known. Another star—one not far away as astronomical distances go—an invisible infrared member of a star class known as Epsilon Aurigae, is so large that it would fill up our entire solar system as far out as the 5.5-billion-mile circumference of Saturn's orbit. This star is just about 100,000,000,000,000,000 times as large as the smallest star thus far discovered. This smallest of stars has a diameter of only 2,500 miles and is actually smaller than some of Jupiter's moons. If it were not for the fact that this tiny celestial body is very dense and glows with its own light, it would hardly qualify as a star at all.

The Epsilon Aurigae giants are sometimes described as "red-hot" vacuums. The matter that makes up these strange stars is thousands of times thinner than ordinary earthly air. Indeed, if you were in a spaceship, it would be possible for you to rocket through one of these giants for weeks on end without ever realizing that you were inside a star.

At the other extreme exist stars whose density measures some 600,000,000,000,000 times greater than that of these giants. These stars are called "white dwarfs" by astronomers. They are so dense that a cubic inch of material from the core of one of them would weigh about 600 tons, or as much as a modern six-story office building. You would

"Let There Be Light" 11

need a powerful jack to raise a tiny particle of dust on the surface of one of these white dwarfs, and even the thinnest wisp of smoke would thump down to the ground like a piece of lead.

There are stars which throb with the definite rhythm of an internal sequence of atomic convulsions. Some of these stars tick like clocks, some flash their messages in long-drawn-out semaphors, some breathe with regular contractions and expansions, whereas others simply change their color and luminosity; some smolder, some fume or smoke or spit, and some explode in a brilliant pyrotechnic display that dazzles the heavens.

Not only can astronomers describe what these distant stars look like, but, even more amazing, they can also tell us what these stars are made of and what their surface and interior temperatures are. They can analyze, quite accurately, the chemical content of their atmospheres, and they can tell exactly what kinds of nuclear reaction cause the individual stars to shine.

Yet, the closest star to earth is more than four and a half light years away. That is, it takes light, traveling at a speed of 186,000 miles per second, four and a half years to reach us after leaving this star. And this is our closest stellar neighbor!

Most stars, however, are much farther away and some of the most distant galaxies seen through our most powerful telescopes are all of two billion light years away. The light that filters down to our telescopes from these galaxies started its journey before any life at all appeared on our earth, at a time when our earth was still a smoldering lump of molten rock and lava.

How, then, is it possible that we can learn so much about the stars and the universe through this solitary and precarious connection? It is possible because of the marvelous properties of light. Light is matter transformed into pure energy, into a wave that carries along in its swift undulations the code of its origin. As scientists discovered ways and means of understanding this code, they also found the means by which they could comprehend the most distant stars and galaxies that telescopes could perceive.

2

THE SOURCE OF LIGHT

The sun is the source of all light on earth, as well as on all the other planets and celestial bodies of our solar system. The light from the moon and the other planets is only reflected sunlight. Some light, of course, does filter down to us from the distant stars; some is produced chemically; some is given off by radioactive atoms. All this light taken together, however, adds up to only an infinitesimal part of the light that pours down on the earth from the sun.

Even the electricity that lights up our cities and homes can be traced directly to the sun. Most electricity is produced through steam generators that are fired by burning either coal or oil. The coal and oil, in turn, are merely storehouses of the sun's energy. This energy was captured geological ages ago by the leaves of living plants and transformed into carbon. The carbon was deposited in the earth in the form of coal and oil. It is the sun's heat, then, stored in coal and oil, that turns our turbines and powers our generators.

Electricity produced by water power is also traceable to the sun, though this connection is more difficult to understand. It is the sun's rays that make the wind and the rain and the snow. Shining down on the surface of the great oceans, the sun's heat causes water to evaporate. This water vapor, moving through the air, condenses and falls down on the earth again in the form of rain or snow. It is carried aloft by wind currents that are produced by the sun's uneven heating of the earth's surface. Rain and snow provide the water for our rivers and streams; this water powers the giant hydroelectric plants that generate electricity.

The sun, through its powerful gravitational field, also holds the earth and the other planets in their orbital paths. It is the center of our solar system. It dominates the heavens around us. So important is the sun to our well-being that it has been worshiped as a god by primitive men all over the world.

Since the beginnings of time, men have wondered about the sun. They have tried to understand this great ball of fire and trace its effects on our lives. One of the earliest records we have about speculation on the sun and its properties comes down to us from a teacher named Empedocles, who lived in the Greek city of Arigentum in 450 B.C.

"Come," he said to his students, so many years ago, "and I shall tell thee first of all the beginnings of the sun and the source from which has sprung all things we now behold—the earth and the billowy sea, the damp vapor and the titan air that binds his circle fast around all things."

Even before Empedocles and the Greek philosophers, men had an intuitive understanding of the sun. They elevated it to the position of a deity and paid homage to

SUN SYMBOLS

it in holy worship. The Egyptians called their sun god Ra and built their giant pyramids in his honor. The Persians worshiped him as Mithras, whereas to the ancient Greeks he was Apollo, who drove a fiery chariot across the vault of heaven every day. Blazing deities from China to Parthia to Stonehenge, in England, gave mythic stature to the sun. The Inca—the New World children of the sun—raised huge pyramids in his honor. And the Japanese emperor was once believed to have been directly descended from the sun, his visage so bright that mortal man dared not stare directly at him for fear of being blinded by the light that emanated from his face.

What, then, is this giant orb of fire? Where did it come from? How can it radiate so much light and energy over such a long period?

These are questions that men have pondered for as long as they could see the light of the sun and feel its heat. Today, however, for the first time, scientists are finding some of the answers. Scientific observation and analysis are solving the ancient mysteries.

We know now that not only is the sun actually a star —a star being a hot, light-making body—but it is the star closest to Earth. We also know that our sun, for all its glory when compared to our dark planet Earth, is quite an ordinary star. It is bigger than some and smaller than others, and it is a member of the largest single class of stars that astronomers have classified. Compared to its planets, however, our sun is truly a giant.

It would take 300,000 earths, for example, to make the sun. Indeed, if all the matter in the solar system—including all the planets, asteroids, satellites, meteors, and cosmic debris that floats within the orbit of Pluto, the planet farthest from the sun—were lumped together, we

The Source of Light

would find that the sun made up 98 per cent of all this matter.

Also, our solar system is flung much farther out into space than the bare figures suggest. When we say that the earth is about 93,000,000 miles away from the sun, it is difficult to imagine what this distance means. To help us get a better idea of the huge size of our solar system, let us imagine a model reduced about five billion times.

In this model, the sun could be represented by an ordinary soccer ball about twelve inches in diameter. Now, how far away from the sun would you imagine the planets to be, on this scale? A few feet, or perhaps several yards? They would be much farther away!

Mercury, as small as a tiny speck of dust on this scale, would be forty-two feet away. Venus, another speck about twice as big as Mercury, would be seventy-eight feet away. Earth, about the size of a pinhead, would be one hundred and eight feet away; Mars, fifty-four yards; Jupiter, one hundred and eighty yards; Saturn, three hundred and forty yards; Neptune, one thousand and eighty yards; and tiny Pluto would be more than a mile away from the sun. And, beyond this model of the solar system reduced five billion times, the nearest star, on the same scale, would be 4000 miles away!

As you can see from this model, the sun and its planets are spaced very far apart from one another, and the stars are spaced even farther apart. The sun controls this system of planets, satellites, and cosmic debris that spreads over an extremely large area. It holds all these stellar bodies within the bond of its powerful field of gravity. But, more than merely holding the planets in their places, the sun also bathes them in an unending flow of radiant energy.

Where does this radiation come from? How is the

sun able to create such vast amounts of light, heat, and energy, whereas the earth and the other planets are cold, dark worlds that can shine only through the reflected light of the sun?

In attempting to answer these questions, let us look first at Earth—the stellar body that we know the most about.

We know that the pressures at the center of the earth are enormous. Bearing down on the earth's center is not only the gravitational force of the earth, but also the immense weight of overlying rocks and atmosphere. Indeed, this pressure is estimated to reach values of about 50,000,000 pounds per square inch in the deep interior of the earth. This fact of pressure holds true for the other planets too, and those that are larger than the earth have correspondingly higher pressures. That ordinary solids and liquids can stand up against such tremendous pressures without collapsing in on themselves is a surprising enough fact; but what must happen to atoms in the heart of the sun which are subjected to much greater pressures?

The sun, as we have already seen, is immensely larger than even the greatest planet. The pressures in the heart of the sun, then, assume truly staggering proportions; they are hundreds of thousands of times greater than those pressures that occur in the heart of the earth. These pressures within the sun are so great that ordinary solids and liquids cannot withstand this tremendous force. Indeed, if the sun were made up the way the earth is, we would see it collapse before our very eyes under the awesome weight of its own gravity.

But the sun does not collapse. It remains stable over long periods. The sun has not changed in size or in the amount of radiation it sends out into space for at least the past 500,000,000 years. This figure has been determined by geologists who study the crust of the earth and early fossils of plants and animals. Their findings show that, during the period since life first appeared on earth, the amount of radiation we have received from the sun has not changed to any considerable degree.

There is only one possible way that the sun could maintain its size and shape over such long periods of time: the material within the core of the sun must be very, very hot. This is the only way the core of the sun could successfully resist the tremendous weight and pressure that squeeze down on it like a giant press.

When matter becomes hot, it expands. It is this property of matter that keeps the sun stable. The expansion of the hot matter in the heart of the sun balances the force of gravity and pressure pushing down on the center.

We also know that, when you apply pressure to matter, this pressure causes the matter to become hot. This process is what makes the material in the center of the sun and the material in the center of the earth hot.

Within the past few years, scientists have developed a method for calculating the temperature in the center of the sun. They have found that this temperature must reach a value of at least 15,000,000 degrees Centigrade. This is so hot as to be almost unimaginable. On earth, for example, the hottest electric ovens can only reach a temperature of some 3,000 degrees Centigrade.

A temperature of 15,000,000 degrees, even one radiating from a body as far away as the sun, would vaporize everything on earth in a matter of seconds. Fortunately for us, the surface of the sun is much cooler than the center

—only about 6,000 degrees Centigrade. If the surface were much hotter than this, neither you nor I would be here to wonder about its temperature.

Where does all this heat come from? Actually, the pressures that occur in the heart of the sun could, in themselves, account for the high temperature. That is, they could account for it providing no heat escaped from the sun. Heat, however, is always pouring out of the sun, radiating from its surface in a steady flow that has not changed in volume or intensity for at least the past half billion years. If the high temperatures in the center of the sun depended only on this internal pressure, it would have cooled off millions of years ago after the heat generated by the pressure had radiated entirely out into space. Thus we conclude that there must be another cause for the sun's heat.

If the sun were made of the best grade of coal or oil mixed with pure oxygen and this mixture were ignited, we could easily calculate exactly how long the coal could burn. For all the immense size of the sun, it could not burn very long, not much more than two or three thousand years. All that would be left of the coal-sun after this period would be a huge chunk of cooling ashes and gas. But we know that the sun has been shining, not for thousands but for billions of years, and it is destined to shine for billions of years to come. We must look for another, much more efficient, energy source for the furnace of the sun.

This problem of the energy source of the sun baffled scientists for centuries. It was not solved until Albert Einstein and other twentieth-century giants of science formulated an entirely new concept of the universe. Einstein's relativity equations offered the first clue to the source of the sun's energy. When he demonstrated the equivalence of energy and matter in the now well known equation

($E = MC^2$), Einstein discovered the sun's elusive secret.

Einstein's great insight into the workings of the universe proved a boon to science. All over the world, scientists pondered his new ideas and applied them to the problems of the laboratory. Two of the most important developments that came from these new theories were the release of the energy potential of the atom and the solution of the problem of the sun's energy source.

Scientists learned that the great solar furnace depended on a nuclear, not a chemical, reaction. "Nuclear fusion reaction" would be the more accurate term, because, like the hydrogen bomb, the sun generates its energy through the fusion of four atoms of hydrogen into one atom of helium.

These nuclear explosions go on continuously within the core of the sun. Because of the immense size of the sun, the tremendous amounts of energy released by the nuclear explosions have struck a balance with the amount of energy that the sun radiates into space. The energy that the sun loses through this constant radiation is compensated by the energy generated through the nuclear reactions.

This balance creates the exact conditions necessary for life to be able to flourish on earth. If the amount of energy leaking through to the sun's surface were greater than the amount that is radiated into space, the sun would simply warm up until the exact balance was struck. In this case, the sun would be too hot to support life on earth as we now know it.

If the energy generated through the nuclear reactions was much smaller, then the sun would cool until a new balance was struck between the energy generated and the energy radiated from the sun. In this case, the sun's rays would not be very strong; earth would then become too cold for life to exist on it. However, a critical balance is maintained by the nuclear reactions that occur in the heart of the sun—the fusion of four hydrogen atoms into one helium atom. Scientists, noting this phenomenon, theorized that, if hydrogen is the fuel that keeps the sun going, the sun must have been made originally of almost pure hydrogen. Observation proved that this was indeed the case. Recent analyses of the sun estimate that it was originally composed of 99 per cent hydrogen, with the remaining 1 per cent being made up of all the other elements.

Taking this idea a step further, we conclude that there must be a great deal of hydrogen spread throughout the universe. Observation has borne out this theory also. Throughout the seeable universe, wherever our giant telescopes have peered, astronomers have found thin clouds of hydrogen. These clouds, however, are extremely thin, containing an average of no more than one atom of hydrogen within the space of, let us say, a ten-gallon can.

At first sight, this would not appear much hydrogen, especially when you consider that a square inch of ordinary air on earth contains hundreds of millions of atoms. But the universe is unimaginably large—so large, in fact, that it has been estimated that, if all of this thin hydrogen that is spread throughout the empty reaches of the universe was brought together, it would exceed in weight all the matter of all of the seeable galaxies. We see that hydrogen is, indeed, the most common element in the universe. And it is not too surprising to learn that our sun and our solar system are made up mainly of this very light, simple gas.

Most astronomers agree that our sun and solar system were formed from the condensation of a giant hydrogen cloud. According to current theory, it happened something like this: first, there were the random atoms of hydrogen floating through the empty reaches of interstellar space. These random atoms, because of their mutual gravitational attraction, tend to clump together into thin, diffuse clouds. These clouds of interstellar gas are quite common and are easily seen throughout our own Milky Way galaxy. A large cloud actually hides the center of the Milky Way from our telescopes. Once the clouds have formed, the gravitational attraction of the individual atoms becomes stronger as they come closer together. The gravity then attracts more atoms and grows steadily thicker. As this condensation compresses the stellar cloud, pressure on the atoms in the center causes them to heat. When the temperature becomes high enough, reaching a value of some 12,000,000 to 15,000,000 degrees Centigrade, energy begins to be generated through the nuclear fusion process.

Eventually, a stage is reached where the energy generated within the new sun strikes a balance with the energy that is radiated into space. At this point, all further contraction stops, and the body of the star becomes stable. To sum up, then, the stages of star birth: first, random atoms that tend to gather into clouds as a result of gravitational attraction; these clouds condense, causing the internal temperatures to rise because of the compressing action of the contraction; finally, a critical temperature is reached that triggers the nuclear reaction, which causes the completed star to radiate heat and light.

Our sun, according to recent estimates, is now about five billion years old. During all these years, our sun has been constantly using its original supply of hydrogen in the nuclear reaction which transforms the hydrogen atoms into helium. We know that this process cannot go on forever; someday it will have to come to an end. Indeed, the sun has already used up almost half its original hydrogen supply.

The sun, therefore, is destined to run out of fuel within the next five billion years.

Oddly enough, as the sun runs out of fuel, it will grow larger and much, much hotter before it finally cools. This occurs because the balance between the hot interior and the cooler surface is breaking down. As more hydrogen is transformed into helium, the energy generated in the heart of the sun will be unable to rise to the surface as easily as it now can. This is because helium is more opaque than hydrogen to the waves of heat, and heat energy will thus not be able to pass through helium as easily as it passes through hydrogen.

The result will be that the energy, unable to escape at its present rate, will cause the pressure in the core to rise. This pressure will increase until the outer layers of the sun can no longer control it. Then the hot center will expand, like whipped cream flowing out of a pressure can, causing the sun to grow larger. As the sun swells, the heat from the interior will erupt through to the surface, making the surface so hot that the earth and everything on it will be completely vaporized by the soaring temperatures.

The sun will change into a huge star. It will grow until it completely fills all the space as far out, possibly, as the orbit of earth. Then it will reach a point where the energy radiated into space will become greater than the energy generated at its center. At this stage, the sun will gradually begin to cool and shrink. Finally, the sun will shrink into an extremely dense, extremely small kind of star, called a white dwarf by astronomers. Long before the sun reaches this final stage in its evolution, all life on earth will have ended.

Actually, this ultimate catastrophe will not come to pass for a long, long time—at least four billion years. This will give man time enough to learn more about the universe and its wonders. Indeed, by that time man will have perfected space travel and may have learned how to travel beyond the sun to find a new haven on a planet in orbit around a younger, friendlier star.

In the meantime, light will remain the same. Men will continue to study this part of the electromagnetic spectrum that our eyes can see. And these mysterious waves of visible radiation hold in them the innermost secrets of matter. Born in the turbulent cauldron of the sun, light offers a key to knowledge—one of the most important keys, as we shall learn.

3

THE SPECTRUM

All of us have seen rainbows. We have marveled at the band of bright colors that appears when sunlight passes through rain falling from a cloudburst. It is a beautiful sight, and it is easy to understand why people have always been fascinated by rainbows.

A favorite story about the rainbow foretells a pot of gold at the end of each one. To become rich, all you need do is follow a rainbow to its end, and there, waiting for you, will be your pot of gold. Of course, as we grow older, we realize that this is just a charming story, that there is no pot of gold at the end of the rainbow. Yet, in a way, this old fable is true. There is wealth at the end of the rainbow, only the wealth is not gold but knowledge, knowledge vaster and richer than any pot of gold could ever be.

The rainbow has always been a source of wonder. In Biblical times, men hailed this natural phenomenon as a sign from God that the earth would never again be completely flooded. This interpretation is not difficult to understand. The rainbow, usually forming after the sun breaks through the clouds on a rainy day, is the herald of the passing of the storm. When you see the rainbow, you can be sure that rain will soon stop falling.

As the rainbow is the herald of the calm after the storm, so is it also the herald of color. And color, as we shall see, is closely tied up with the very heart of matter. It would be almost impossible to truly understand the universe without first understanding the properties of color. But, even without color's intimate connection with matter, it would still be an important part of our lives. A world without color would be a poor and dull place. It would be like a scene lit by the moon. Everything would show only in variations of gray. There would be no reds or yellows or greens to delight our eyes.

This fact reveals one of the most interesting peculiarities of light: long ago, men noticed that the ability to see color fades sooner than the ability to see light. The light of the moon, for example, is bright enough to see by, but it is not bright enough to reveal colors. The moon's light strikes the earth's surface with only the power of one and one-half candles per square inch. The sun, on the other hand, shines on the same square inch with the power of more than 1,000,000 candles on a bright day. Moonlight, though it is bright enough to see by, is not bright enough to make colors visible.

Color can exist only where there is strong light. Without light, everything is black. In the dark of a closed room where there is only enough light to see by, a red dress is not red, a book jacket is not green, and your necktie is only a shade of gray, no matter how bright its colors may be in the sunlight.

Actually, colors are not really colors at all. Ordinary

sunlight, as is shown by the rainbow, consists of all the colors of the spectrum. When we say that a thing is red, what we really mean is that this material has the ability to use light in a way that allows only that particular color to be reflected back to our eyes. To understand this interesting property of color, we must first understand something of the nature of light itself.

Light is a form of energy. It can travel through space at a fantastic speed, but it cannot be seen until it strikes either the eye or an object which reflects it to the eye. You can easily understand this fact simply by looking around you. Light from a source—from the electric bulb that you may be reading this book by, for example—spreads out in all directions until it is stopped by objects in the room. Then, from each of the multiple surfaces on these objects, the light is reflected into your eyes, bringing an image to your brain. Imagine, if you can, the resulting confusion if your eye could also see each beam of light as it traveled about the room from the lamp to the object and from the object to your eye. If this were so, your room would appear as a mad jumble of crisscrossing beams of light. What chaos there would be!

Fortunately, light cannot be seen as it moves; however, this does not mean that light is immaterial, a kind of nothingness. Light is very real. It exerts a pressure that can be measured and felt. The real pressure of sunlight, for example, is what causes the tail of a comet to point away from the sun no matter what its orbital position might be.

This very real pressure was recently demonstrated by Echo I, the giant communication satellite that could be seen with the naked eye as it moved around the earth. Echo I was pushed out of its calculated orbit by the steady pressure and weight of sunlight beating on its surface. Indeed, this pressure proved to be so strong that today scientists are developing "solar sailing ships" to utilize this force to propel them on interplanetary trips. These space ships will be equipped with giant solar sails that will catch the rays of light from the sun the way a sailboat catches the force of the wind. Because light travels so much faster than wind, scientists are hopeful that their solar sailing ships can attain the speeds necessary to move across the vast gulfs of space that separate the planets.

The question of how light travels has puzzled men for hundreds and thousands of years. At one time scientists believed that all space was filled solidly with something they called the "ether." This concept of an all-pervading medium helped scientists explain many things.

In 1637, René Descartes, the great French mathematician who is considered the father of modern mathematics, proposed the theory that light at its source exerts a pressure on the ether. This pressure, in turn, causes the ether to press against the eye, thereby creating the sensation of light. It was, in Descartes' view, like pressing a long rod against the wall. When you push one end, the other end exerts pressure on the wall. To account for color, Descartes reasoned that the pressure might also be given a turning motion. The amount of turn determined the color, red resulting from the fastest twist, violet from the slowest.

Descartes, however, was only one of many distinguished scientists who were working with light and color during the 1600's. This interest was sparked when a study of light by the Arabian mathematician and scientist, Alhazen, was translated into Latin and made available to European scientists. This book was passed on from country to country and inspired many thinkers to the study of the unique properties of light.

René Descartes (1596-1650)

Before the publication of Alhazen's work, most European scientists accepted Aristotle's theories almost without question, so great was the reputation of the Greek philosopher. Aristotle, who lived in the fourth century B.C., theorized that white light was pure—fundamental and perfect. Darkness, on the other hand, was a property of imperfect earthly matter, an impurity that affected the purity of white light. When this heavenly purity became infected or mixed with earthly darkness, color resulted. A small amount of earthly infection, according to Aristotle, produced the color red, greater amounts produced blue, and so on until all the colors of the spectrum were finally produced.

Alhazen, however, proposed a new theory, one which was daring and radical according to the standards of the time. The Arabian savant said that light spread out in all directions from each point on a visible object in the form of a continuous stream of tiny globules. Color, according to his theory, was the result of greater or smaller concentrations of these tiny particles of light. This theory, of course, was very different from the accepted theory of Aristotle. It was exactly this difference, which challenged Aristotle's cherished notions, that sparked the great interest in light.

Descartes proposed his theory after having studied the work of Alhazen, but other theories soon appeared. Another scientist who read the Arabian's study was an Italian priest and physicist named Francesco Maria Grimaldi. After reading Alhazen's book, Grimaldi began experimenting with lights and shadows. Out of his observations came an entirely different theory. Light, he reasoned, actually moved very swiftly through the ether. It traveled in the form of waves, in much the same way as a ripple travels through water when you throw a rock into a pool.

In England, at about the same time, a brilliant scientist with a terrible temper was also at work on light. Robert Hooke was born on the Isle of Wight, off the coast of England. His father had hoped that his son would some day enter the ministry. Robert, however, was very sickly as a child, and, after he had suffered an attack of smallpox that left him terribly scarred, his father abandoned his education. He felt that the boy would never be healthy enough to successfully pursue a career as a priest.

Left to himself, young Robert developed a profound interest in mechanics and mathematics. He built and designed elaborate mechanical toys and models. After his father's death, Robert was sent to school in England by an uncle. He amazed his teachers with his ability in mathematics. On the strength of his astonishing ability, he was admitted to Oxford University and worked his way through school by waiting on tables and working as a servant.

His illness, the ugly appearance which was a result of the smallpox, and his poverty did nothing to improve his temperament. Instead, they seem to have combined to produce an overwhelming and touchy sense of pride which made it hard for anyone to work with him. Hooke was always ready to take offense and to defend himself against even the slightest of imagined insults. His entire career was marred by petty bickerings and long-drawn-out quarrels with his colleagues. But, despite his bad temper, Robert Hooke's genius won him the respect of the English scientific world of his day. He was appointed secretary of the Royal Society, a group which counted among its members the outstanding men of science in England.

In 1655, Hooke published *Micrographia*, a book in which he developed the theories first put forth by the Italian, Grimaldi. Hooke agreed with the friar and said that light travels in the form of small, very rapid vibrations. Colors, according to this theory, were produced by variations in these vibrations.

In Holland, across the channel from England, another scientist, named Christiaan Huygens, read Hooke's book and was deeply influenced by the theories of the irascible Englishman. Huygens developed these theories still fur-

The Spectrum

ther. He postulated light as "a successive movement, impressed on the intervening ether."

Through this process, described in his *Treatise on Light*, Huygens showed that light spreads "as sound does, by spherical surfaces and waves: for I call them waves from their resemblance to those movements formed in water when a stone is thrown into it and which present a successive spreading of circles." Huygens brilliantly developed the wave theory of light into the important concept that is still known as Huygens' Principle. He showed that each point on any advancing wave front is in effect the source of a fresh wave, and all such fresh waves continue together as the advancing front. This principle explains why a loud sound originating outside your open window, for example, produces something like a new sound source at the window.

The theories of Hooke and Huygens represented a bold but reasonable attempt to explain one of the most puzzling phenomena of nature. But the work of these two men was soon to be eclipsed by that of one of the true giants of science, Sir Isaac Newton. During these same fateful years, he, too, turned the power of his intellect to the problems of light.

Soon after graduating from college, while still a young man, Newton busied himself with the construction of a telescope. Though he worked very carefully and precisely, he found that, no matter how hard he tried, a faint ring of colors always blurred the outside edges of each lens. This persistent color ring fascinated Newton, and he decided that he would have to know more about light and color before he could go on with his telescope construction. He stopped grinding lenses and began, instead, to study the properties of light.

This work resulted in the first scientific paper that he

presented to the Royal Society—an account of "the oddest, if not the most considerable detection, which had hitherto been made in the operations of nature."

In 1666, Isaac Newton wrote in this report:

> I procured me a triangular glass prism, to try herewith the celebrated phenomena of colors. And in order thereto, having darkened my chamber, and made a small hole in my window-shuts, to let in a convenient quantity of the sun's light, I placed my prism at its entrance, that it might thereby be refracted to the opposite wall.

"It was at first a very pleasing divertisement to view the vivid and intense colours produced thereby," Newton touchingly admitted.

The colored band that spread onto Newton's wall, ranging from red at the bottom to violet at the top, had more of a scientific than an esthetic interest for the scientist. He applied himself "circumspectly" to the problem of exactly what made the light separate into these colors. Newton continued:

> Then I began to suspect whether the rays after their trajection through the prism, did not move in curve lines, and according to their lesser or greater curvity tend to divers parts of the wall. I remembered that I had often seen a tennis ball struck with an oblique racket, describe such a curve line. For, a circular as well as a progressive motion being communicated to the ball by the stroke, its parts on that side where the motions conspire must press and beat the contiguous air more violently than on the other, and there excite a reluctancy and reaction of the air proportionately greater.

In this treatise on light, Newton also incidentally explained the principle of the curve ball as it is pitched in baseball today. By spinning the ball as it is thrown, the pitcher causes the ball to move toward the batter in a curved path, exactly as described by Newton some three hundred years ago. The sharpness of the curve is determined by the amount of spin and the speed of the pitched ball.

This line of reasoning led Newton to the theory that light, too, must travel in the form of tiny balls. He explained it this way:

> And for the same reason, if the rays of light would possibly be globular bodies, and by their oblique passage through one medium into another, acquire a circular motion, they ought to feel the greater resistance on the ambient ether on that side where the motions conspire, and thence be continually bowed to the other.

Newton explained colors as the components of pure, white light. Light, when it passed through a prism, was really being sorted out into its fundamentally different "rays, some of which are more refrangible [bendable] than others."

Red, according to Newton's theory, was the least "refrangible," whereas violet, which fell on top of the spectrum, was the most. "The most surprising and wonderful composition of all," Newton wrote, "was that of whiteness." White could be made by remixing all the colors thrown on the wall by the prism by passing them through a second prism. This observation confirmed the fact that ordinary white sunlight was compounded of many colors.

From this, Newton correctly concluded that "colours are not qualifications of light derived from reflections of natural bodies, as 'tis generally believed, but original and connate-properties, which in divers rays are divers." Newton's theory, which he called the "corpuscular theory of light"—describing light as tiny, swift-moving globes—was, of course, completely opposed to the wave theory of light championed by Robert Hooke.

Since Hooke was, at the time, secretary of the Royal Society, Newton's paper came to his attention. Because the theories presented were at odds with his own ideas on the subject, Hooke attacked the paper violently. He seemed to have taken it as a personal insult, and, in his rebuttal, Hooke displayed the pettiness and touchiness that so marred his character.

Newton was somewhat taken aback by the violence of Hooke's attack and defended his views in a series of reasonable and carefully thought out letters. Other scientists soon entered the argument, and Hooke was backed by Huygens, who was the outstanding supporter of the wave theory. The arguments became so emotional that, at one point, Newton became so annoyed that he vowed never again to publish any other scientific work.

The bitter debate raged for years. At first, Hooke and Huygens received most of the support. Later, after Newton had relented his vow and let his monumental work on gravity be published, he became famous, and the tide of opinion turned. People believed that it was Newton who could do no wrong. His corpuscular theory of light was re-examined and accepted almost as completely as the theories of Aristotle were accepted a scant hundred years earlier. The work of Hooke and Huygens was ignored, and the wave theory was discredited in favor of Newton's ideas.

Then, science performed another of its famous about-faces. In the early part of the nineteenth century, the experiments of the French physicist, Augustin Fresnel, showed conclusively that the known phenomenon of light could best be explained by wave theory. Science switched sides again, complaining that Newton had delayed progress for a hundred years by overshadowing the work of Hooke and Huygens.

Newton's corpuscular theory, however, proved to be sturdier than its opponents suspected. New defenders appeared to question the findings of Fresnel, and the opposing theories continued to compete for acceptance for another hundred years.

Today, oddly enough, the generally accepted theory of light combines the ideas of both these seventeenth-century views. Our current theory is called the "quantum theory," and developed from the work of such twentieth-century geniuses as Albert Einstein and Max Planck.

The quantum theory assumes that light is given off from a source as separate bundles of energy. Each bundle travels in a fixed pattern, or wave formation. These tiny bundles of light—or "quanta" of energy, as they are called by scientists—are given off at such a rapid rate that there are no sizable gaps between the individual bundles. Thus, quantum theory states that light is both a wave and a corpuscle and combines the competing ideas of the two Englishmen.

The work of Newton and Hooke, however, provided a giant step toward the fuller understanding of light. The bitterness that the two theories generated was caused by the the fact that, as modern science has discovered, both theories were partially right. Thus it was that these two enemies placed valuable tools in the hands of their successors. Using Newton's experimental methods and Hooke's and Huygens' theories, great progress was made in many areas toward the fuller understanding of the universe.

Newton had used the word "spectrum" to describe the

colors which combine to make white light. Now other scientists began to apply the wave theory in their study of the colored lights of the spectrum. The beautifully glowing bands of color that so fascinated Newton were to prove a treasure store of information, far richer than the early imaginings of either Newton or Hooke.

4

PRACTICAL APPLICATIONS

While the theorists theorized and the learned physicists and mathematicians debated bitterly over the true interpretation of light, other, more practical, men went unheedingly ahead with the task of putting light to practical use. While Newton and Hooke fumed and the scientific world joined the speculation as to whether light was made up of tiny corpuscles or immaterial waves, the lensmakers made giant strides in the perfection of both the microscope and the telescope. Though these men did not know exactly what light was, they knew how it behaved when it passed through a lens. Their work brought the distant stars closer to man's eye and made the invisible world of the microcosm visible.

Actually, men had learned to manipulate light—to bend its rays by means of specially shaped bits of glass—very early in history. Archaeologists, for example, unearthed a lens fashioned from rock crystal in the ancient city of

Nineveh. This lens was made at least four thousand years ago.

The ancient Greeks were also well acquainted with lenses. They ground them out of crystal and glass, and they made hollow glass spheres that they filled with water. Many of these spheres have been discovered among Greek artifacts. Water, of course, can make a very effective lens, as you can see for yourself by holding a glass filled with water in front of this page. The letters directly behind the glass will be magnified.

The ability of water to bend rays of light was noticed long before the Greeks made their hollow glass spheres. The very first man, squatting at the edge of a clear stream, who reached into the water to pick up a pebble must have noticed this property. The pebble was not exactly where he "saw" it. Something happened to the rays of light as they passed from the pebble up through the water and into his eyes. This something is called "refraction." It is the change in direction of light when it passes from one medium into another.

This property was carefully studied by Ptolemy of Alexandria during the first century A.D. In his famous work entitled *Optics*, Ptolemy compiled a complete table showing the exact amount of refraction (bending) of light by various angles of incidence as it passed through glass, water, and other transparent materials. Ptolemy computed his table by carefully measuring light rays as they passed through these various mediums at varying angles.

According to Ptolemy's findings, light moves at varying speeds through various mediums. It moves quickest in the air, he discovered, and slows somewhat when it passes through glass. It slows even more when it passes through crystal or water. Ptolemy explained refraction by assuming that light traveled as a beam. When the beam entered a lens or water straight down, the entire beam was slowed evenly. Because of this even change of speed, the beam is not bent as it passes through the medium, and no change in the direction of the light can be observed.

When light strikes the new medium at an angle, however, the beam does not enter all at once. Part of it goes in ahead of the rest of the light beam, and this part is slowed while the rest of the beam is still traveling at the original speed. This uneven slowing of the beam causes it to be bent. The sharper the angle of incidence, the more the beam is refracted.

The refraction of light explains many strange things. It explains, for example, why a pebble under the water is not where you "see" it to be. It also explains how mirages occur. Light can also be bent as it moves through layers of air that have different temperatures. It moves faster through warm air than it does through cool. It is this property of light that produces mirages in the desert or at sea. When you see a "ghost ship" sailing in the clouds at sea, what you are actually seeing is a real ship that is somewhere beyond the horizon. Immediately above the surface of the sea, there is a layer of cool air. Above this layer, the air might be warmer. Light reflected from the ship beyond the horizon is bent as it passes from one layer of air to the other. Your eyes see the refracted light from the ship as though it had traveled in a straight line, thus producing the mirage.

By carefully measuring exactly how much light was bent by various angles of incidence and by various mediums, Ptolemy compiled his tables. The accuracy of his figures, as revealed by comparison with modern measurements, testify to the care with which this great astronomer worked.

Although his surprisingly advanced theories could hardly have been widely appreciated at the time, they do explain how accurate lenses came to be used in Rome for what Seneca described as the "magnification of writing."

Most of these early lenses, however, were crude affairs. Ground crystals and spheres of water were used mainly as novelties, entertaining the ancients with their distorted images of everyday things. Although optic properties were tabulated by Ptolemy and others, practical use of the lens for magnifying distant or very small objects did not come about for another thousand years.

Glasses, as we know and use them today, did not come into general use until sometime around A.D. 1200. By 1300, however, they had become quite common. About this time, the glassworkers guild of Venice listed *roidi da ogli*—"small disks for the eyes"—among the objects that guild members were qualified to manufacture. For a time, the techniques for manufacturing these "disks for the eyes" was a jealously guarded secret, and the Venetian glassmakers enjoyed a monopoly, exporting their glasses to all of Europe.

A painting of a Dominican monk dated 1352 shows how these Venetian "disks for the eyes" were used. The monk is wearing what appear to be two small hand magnifying glasses with their handles joined in a hinge. The glasses are held in place by squeezing the two lenses against the bridge of the nose. The monk, obviously a scholar, is peering through them at a manuscript with fine writing on its pages.

The Venetian monopoly did not last long. We know that, by the beginning of the fifteenth century, the art of lensmaking had spread throughout Europe. Practically every city on the Continent had its guild of glass workers who fashioned the crude spectacles and magnifying glasses of the time. Looking back now, we cannot help but wonder why the telescope did not quickly follow the eyeglass. Both, of course, work on the same principle, and today it seems obvious to us that a lens that could make things look bigger could also make distant objects seem closer.

It seems obvious to us now, but apparently it was not then. Ptolemy, who made such careful optical measurements, never thought to use his lenses to look at the stars. Not even so universal a thinker as Leonardo da Vinci, who wore spectacles in his old age, could come up with this development. Nothing even closely resembling a telescope appeared until about 1590.

Like so many other inventions, that of the telescope has been claimed by many countries. Giambattista della Porta, an Italian physicist who divided his time between science and writing comedies for the theater, mentioned using a combination of lenses to make distant objects appear near. England also claimed the honor of having invented the first telescope by virtue of the work of two mathematicians, Leonard Briggs and John Dee, who experimented with lenses in the late sixteenth century.

It is a Dutchman, however, named Hans Lippershey who is credited with the invention of the first practical telescope. Lippershey was an ordinary spectacle-maker who lived in Middleburg, Holland. Quite by chance, he discovered the effect of combining a convex and concave lens. As legend has it, his children made the actual discovery while playing with some of his lenses. However it came about, it must have seemed like a miracle to the spectacle-maker. Suddenly, with the aid of the lenses, a ship far out at sea appeared to be just outside his window. The discovery was bound to be made.

Hans Lippershey may have been an ordinary spectacle-maker, but he was an excellent businessman. He saw the practical value of his telescope immediately, especially as a military aid. Lippershey mounted his lenses in a long tube and offered his invention—at a price—to the Dutch government, which at the time was in the midst of a prolonged war with Spain. Dutch naval officials tested the telescope and were enthusiastic in their findings. They suggested to Lippershey, however, that he make a telescope that could be used with both eyes.

Lippershey attached two of his telescopes and made the first pair of binoculars, which became famous throughout Europe as "Dutch Trunks." Today, only the carrying-case and a few technical refinements represent the difference between modern binoculars and Lippershey's original invention. The lens arrangement—the most vital part—is the same. A concave lens is placed in the eyepiece, and a convex lens is placed at the other end of the tube; the calculated refraction of light causes distant objects to appear close.

News of the Dutch spectacle-maker's invention spread throughout Europe and reached the attention of Galileo Galilei in Italy. He wrote in his journal in 1609: "A report reached my ears that a Dutchman had constructed a telescope, and some proofs of its most wonderful performance were described." Galileo was fascinated by the idea of bringing distant objects closer. He learned all he could about Lippershey's seeing aid, and, after having spent some months in "deep study of the principles of refraction," he prepared a lead tube about $9\frac{1}{2}$ feet long by $1\frac{2}{3}$ inches in diameter. To the ends of this tube Galileo fitted "two lenses, both plane on one side, but on the other side one spherically convex and the other concave. Then bringing my eye to the concave lens I saw objects satisfactorily large and near, for they appeared one third their true distance off and nine times larger than natural."

Practical Applications 45

Though this was the most powerful telescope ever made up to that time, Galileo was still unsatisfied with its performance. During the next few months, he built a second telescope, "with more nicety, which magnified objects above sixty times," and a third that was still more powerful. With this instrument, Galileo "betook myself to observations of the heavenly bodies . . . with incredible delight."

Like Lippershey, Galileo also saw the military possibilities of the telescope. He presented one of his first to

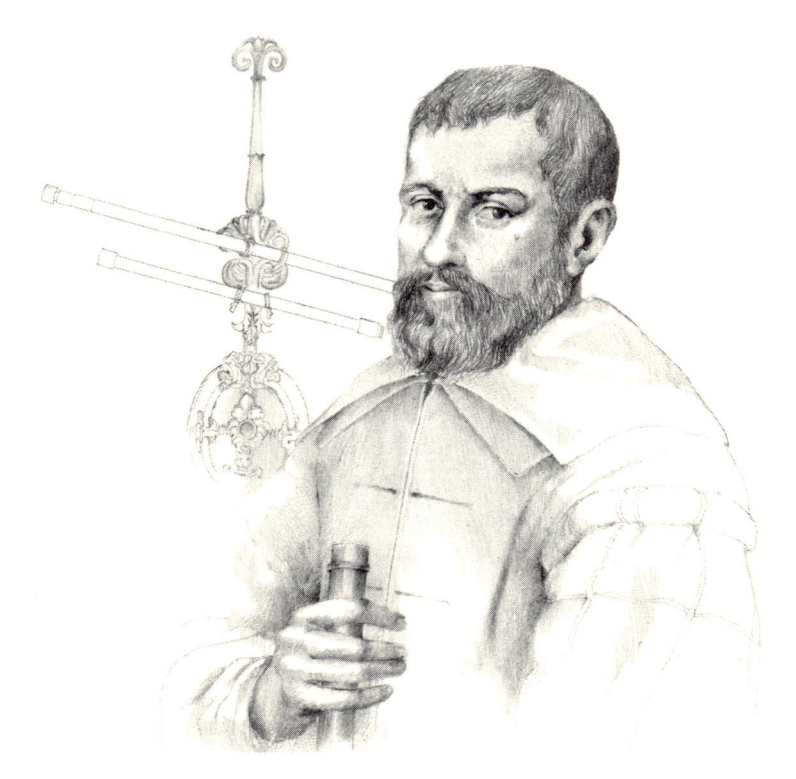

GALILEO GALILEI (1564-1642)

the government in Venice. So pleased were the naval authorities with its performance that Galileo was rewarded with a lifetime appointment to his chair at the University of Padua at an increased salary. For himself, however, Galileo preferred spying on the heavens, and every clear night saw him standing before the telescope on his roof peering at the stars.

During his first few months of observation, Galileo discovered hitherto-unsuspected dimensions of the universe. He could hardly contain his delight. The moon's face, for example, "was full of hollows and protuberances." The planets "present their disks round as so many little moons." He discovered the four "sidereal bodies performing their revolutions around Jupiter . . . the four Medicean satellites, never seen from the very beginnings of the world up to our times."

Galileo first saw the rings of Saturn and found, in every direction that he turned his telescope, that there were "stars so numerous as to be almost beyond belief." He catalogued at least eighty new stars close to Orion's belt and forty unsuspected stars in the Pleiades, and the Milky Way was revealed to his eye as "nothing else but a mass of innumerable stars . . . the number of small ones quite beyond determination."

Galileo's pioneer work was followed by that of a long list of distinguished scientists who devoted themselves to the task of making the far-away seem near. Johannes Kepler, the seventeenth-century astronomer and mathematician, was the first man to use two convex lenses to enlarge the field of vision of his telescopes. Kepler's system of lenses was a big improvement and was used thereafter by most telescope-makers. Refracting telescopes, which are based on the work of Galileo and Kepler, have made and are still making important contributions to our knowledge about the universe.

Isaac Newton, the great English genius, also applied himself to the construction and design of telescopes. As we have already seen, in Chapter 3, Newton was dissatisfied with the refracting telescope. The edges of the lens were always blurred with a faint ring of colors. As a result of his study of the nature of light, Newton concluded that this defect, which is called "chromatic aberration," could never be entirely corrected. He searched for some other means with which to bring the stars closer.

At the same time, another Englishman, James Gregory, was also experimenting with light. He discovered that carefully shaped mirrors could be made to do everything that a lens could do. A mirror could also be made to gather light and focus it into a single point. The mirror does this by reflecting the light, rather than bending it the way a lens does. The performance, according to the findings of James Gregory, was the same. He concluded from this fact that a telescope could be made with mirrors as well as with lenses.

Newton learned of the work of his countryman and studied the reflecting qualities of mirrors. From these experiments, he found that a telescope made with mirrors would solve the problem of chromatic aberration. With this fact in mind, Newton proceeded to design and build the first reflecting telescope.

This instrument was only six inches in length, and the open end of the tube was just a little more than an inch wide. The mirror Newton used was made of metal which he had cast and polished himself. Another, much smaller, mirror was inserted at the focus of the larger mirror and reflected the focused image through an eyepiece lens.

When Newton pointed his new telescope at the stars, his hopes were realized far beyond his expectations. The tiny model proved to be as powerful as a six-foot refracting telescope. Best of all, however, there was no chromatic aberration. The image was unclouded by any ring of colors. Newton's telescope amazed the scientific world of his day and helped establish the young man's reputation as an outstanding scientist. The success of his telescope was also an important contributing factor in the widespread acceptance of his corpuscular theory of light.

Today, the largest telescope in the world—the 200-inch Hale telescope of the Mount Palomar Observatory in California—is a direct descendant of Newton's tiny model. The Hale telescope also uses a concave mirror to gather the light from the stars. But, whereas Newton's model had a mirror that measured a little more than an inch, the Hale mirror measures 200 inches across. This huge mirror is made of Pyrex glass and was cast at the Corning Glass Works in Corning, New York, in 1934. It used sixty-five tons of Pyrex that was heated to a temperature of 2,835 degrees Fahrenheit in order to make it molten. The heating process took sixteen days. Then the mirror was poured into a mold with huge ladles that held 700 pounds of molten Pyrex.

To cool the mirror without having the glass crack, it was necessary to lower the temperature very gradually. This cooling process took ten months. Finally, after it cooled without cracking, the huge disk of glass was transported to California, where it was ground to the precise focus. To do this, more than five tons of glass had to be ground out of the disk. This rough grinding was followed by further finishing and polishing until, finally, an aluminum coating was applied to reflect the light.

Next, the giant mirror was mounted in the telescope at the top of Mount Palomar. This huge instrument, that weighs an estimated 200 tons, was as carefully constructed and as accurate as the most delicate Swiss watch. Setting up the Hale telescope was a long, tedious, expensive job, but there is not an astronomer in the world who will not agree that it was well worth all the effort and cost. The 200-inch mirror can gather in the light of stars that were beyond the reach of lesser instruments. Its powerful eye has enabled astronomers to prepare entirely new charts of the heavens.

By utilizing a specially designed camera and color films, the Hale telescope is able to show the stars in full color for the first time. Astronomers have long known that the nebulae, the star clusters that spread throughout the seeable universe, which appear in shades of white and gray through ordinary telescopes, were actually highly colorful. But color, as we have already seen, needs a strong light before it can be discerned by the eye. Lesser telescopes simply could not gather enough light to resolve their actual colors.

Today, the true colors of the stars may be recorded on photographs. We can see, for example, the beautiful filmy scarf of the Veil Nebulae in the constellation Cygnus. The nebulous streaming clouds that always appeared in shades of gray are actually streamers of shimmering reds, whites, and rich blues.

Let us now see how the telescope accomplishes its important task of bringing the stars closer to man's eyes.

5

THE WAY TO THE STARS

With the invention of the telescope and the perfection of more powerful instruments, astronomers had a tool with which to study the heavens in greater detail. From their crude beginnings, telescopes quickly developed. They became larger, more accurate, and more sophisticated. The science of astronomy took giant strides forward. Our bridge to the stars, those waves of light that filter down to the earth, became a broader and more solid structure.

Before we go any further, let us see how a telescope works. Let us see what happens when light is either refracted (bent) by a lens or reflected from a mirror. All of us are familiar with what we might call the everyday language of light. We know, for example, that a light bulb hanging in the middle of a room gives off rays of light in every direction. We can see it from above or below and from all sides as long as nothing interferes with our line of vision. The light from the bulb spreads in rays that travel in a straight line from the source. You can easily prove this fact by placing a book in the path of the light and observing its shadow. The shadow will always mark a straight line from the source of light.

Other qualities of light, however, are more difficult to understand. In your room, for example, it is difficult to prove that the amount of light that reaches your eye becomes fainter the farther you move from its source. Outside your house, where larger distances can be seen, you can easily demonstrate that this is so—that light does become fainter as you move farther from the source.

Look at a street light close by and then look at the same light again from the far end of your block. The light as seen from the end of your block will not seem so bright as it did when you looked at it close by. You could, through careful measurements taken with a light meter, determine exactly how much fainter the light becomes with distance. The same result could be shown more easily by this little experiment.

Take a black sheet of paper and punch a small hole in the center. Now hold it up in front of the light. When the distance between the hole and the source of light is doubled, only one-fourth as much light shines through the hole. In the same way, when we look at a street lamp from a distance of ten feet and then from a distance of one hundred feet, only 1/100 as much light from the same lamp will reach our eyes.

Actually, this is not entirely correct because the pupils of our eyes contract and expand automatically as the light shining into the eyes becomes dimmer or brighter. You can easily see this happen by shining a flashlight into the eyes of a friend. The pupils of his eyes will become larger or smaller as you move the light farther from or closer to his line of vision. A telescope, however, does not have a pupil that changes automatically. If we look through the same telescope at two equally bright stars, one of them twice as far away from the earth as the other, the closer star will cast four times as much light into our telescope. This simple law of light propagation is, as we shall see, extremely useful to astronomers.

Now, let us see what happens when light enters a telescope to be either refracted or reflected, according to the kind of instrument we are using. Most people would say that we use telescopes because they allow us to see distant objects enlarged. Though this idea is not exactly false, it does not really strike at the heart of the question.

What do you do when you want to examine something more carefully? Like most of us, you will probably move the object closer to your eyes. When you do this, you see the object at a larger angle of vision. In this way, the image that enters your eye will cover as much of your retina (the light-sensitive nerve-endings in the back of your eyes) as possible. Of course, you cannot keep bringing it closer indefinitely. You will soon reach a point that is beyond the limit of distinct vision (about ten inches for most of us) and will no longer be able to see the object clearly.

We can, however, bring our eye closer to the object with the aid of a magnifying glass. By bending the rays of light reflected from the object, we see, in effect, not the object itself, but an image of the object. This image has the advantage of being properly placed in relation to the eye. We perceive it at a broad angle while remaining within the limit of distinct vision. The object itself is not changed at all. It is neither bigger nor smaller; it is merely placed in the best viewing position.

Now, how does this property of a lens help us with

our visual exploration of the stars? The difficulty with the stars is that we can neither move them closer to our eyes nor move our eyes closer to them. So, what we have to do is bring an image of the stars into the proper relationship with our eyes. This is the task of the telescope. The principal part of the telescope is the lens, or, as is more generally used, a whole system of lenses that work together. This lens system projects images that actually exist, "real images" of stars that are very far away. In the same way, every camera can be thought of as a small telescope. It captures a real image on the photographic plate—the same image that we can see in the ground-glass viewer of some cameras.

To see exactly how this happens, try this experiment. Get an ordinary hand magnifying glass on a sunny day and hold it out where it can intercept a ray of sunlight. Beneath the glass, the diffuse rays of sunlight will converge into a single, extremely bright point of light. This point becomes so hot that it can burn through a piece of paper or even start a fire. Now, if we move the glass up, the tiny point of light begins to spread; move the glass down, and the circle of light converges again into the bright point. If you keep moving the glass down, the point of light will again begin to spread.

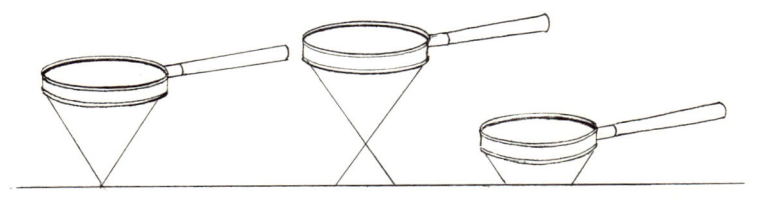

The very bright point of light is called the "focus" of the lens. The light rays from the sun, bending as they move through the lens, come together at this point. Since every point in a real object—rays of sunlight in this case—is reproduced at the focus, we say that a real image is produced. The heat of the focus in this experiment proves the reality of the image.

What happens in a telescope is that this real image is brought closer to the eye. The real image takes the place of the real object. In the case of a telescope pointed at the moon, the lenses bring the real image closer. The lenses do not really "magnify" the moon. The moon remains the same, no matter how powerful the telescope we use. The real image is brought into a focus under which our eye can examine it at a wider angle, and it thus appears larger.

When this real image is impressed on a photographic plate, we have a picture of the moon. We have a permanent record of it in black and white. We can take it home and examine it in still greater detail with a magnifying glass or, better still, with a microscope. We can measure it with a ruler and determine the distances between one point on the picture and another. This is very convenient and, if the measurements are made with a microscopic scale, much more accurate than any measurements made by simply looking could ever be.

Measurements can, however, be made even without a photograph. Standing behind the focus, where the bundle of light rays begins to spread out, we see the image of the moon exactly as though it were on a photographic plate. By using a magnifying glass, we can come even closer to it so as to see it still larger.

When we construct our magnifying glass or a system of several lenses into the telescope in the form of an eyepiece close behind the focus, we have a "visual" telescope.

We can add to its usefulness by introducing various devices —a micrometer or a spectrometer, to name two—close to the focal plane. When the eyepiece is replaced by a camera, we have a "photographic" telescope. Both types of instrument play an important role in astronomical research. What this all means very simply, is that the telescope allows us to see distant objects larger and with greater precision. Through the telescope, the sun and moon appear considerably larger than they do to the naked eye, and the planets, which appear to be no different from the fixed stars when seen without a telescope, become clearly recognizable disks.

If lenses did not have the power to produce real images, we could get no clearer impression of the stars through them than we get with the unaided eye. We cannot, after all, really bring the stars closer. But, since we can bring the pupil of our eye close to the focal point where all the light gathered by the object lens comes together, we are able to take a much larger amount of light into our eye. The larger the opening of the telescope, the more light will reach our eye—or photographic plate, as the case may be—and the brighter the heavenly bodies will appear. For this reason, we build larger and larger telescopes with the largest possible objective lenses.

The use of reflecting telescopes is preferable, since they have an advantage over refractors. Everything that lenses can do with light mirrors can do just as well, as was demonstrated by Isaac Newton some three hundred years ago. In addition, the single surface of the mirror is much easier to grind than the minimum of four optical surfaces required for the lenses of the great refractors. Also, glass in the mirror need not be perfectly free from flaws, since only the reflecting surface really counts. For these reasons, reflecting telescopes are the largest in the world. There are reflectors with mirrors measuring up to 200 inches (the previously mentioned Hale telescope of the Mount Palomar Observatory in California being one), whereas the forty-inch Yerkes refractor, built in 1900, remains the largest of its kind in the world.

Isaac Newton (1642-1727)

A further advantage of the reflecting telescope is that the coating of aluminum or silver with which the concave surface of the mirror is coated reflects practically all the light that falls on it. In and between the great lenses of the refractors, as much as 40 per cent of the light is lost by absorption and reflection.

However, the reflecting telescopes also have their weakness. They provide sharp images of objects only in the center of their field of vision. In sky photographs made with reflectors, the stars are sharp points only in the center of the photographs. Toward the edges, the images tend to become blurred and stretched out. In celestial photos which seek to measure very distant galaxies, this disadvantage is serious, greatly limiting the usefulness of the photographs.

About thirty years ago, however, a German astronomer named Schmidt discovered a method for partially overcoming this weakness in the reflecting telescope. He found that, by placing a very thin, specially ground lens in front of the mirror, a sharp image is obtained even at some distance from the center of the field of vision.

The Schmidt telescope has proved extremely useful. Photographs of stars made with this type of telescope show up amazingly sharp to the very edges of the plate. Schmidt telescopes are, however, twice as long as ordinary reflecting telescopes of the same focal length. The thin lens, called the correcting plate, must be placed at a considerable distance from the reflecting mirror in order to be effective.

The photographic plate or film also has to be curved, so that the image falls on a suitable spherical surface. Despite these two drawbacks, the sharpness of the images obtained with the Schmidt telescope represents such a great advantage that instruments of this type are constantly being built.

The world's largest Schmidt telescope is also at the Mount Palomar Observatory in California. Its mirror has a diameter of seventy-two inches, and its aperture, the opening through which the light enters the instrument, is forty-eight inches. It is the indispensable companion to the giant 200-inch Hale telescope, because the large celestial field it can photograph all at once enables it to pick out the objects which its "big brother" can then examine in detail.

Telescopes, then, are also classified into visual and photographic types. Both are still in use, although the photographic method of observation is the main one today. Photographs have many advantages. They are available at all times and can be examined and studied as often as we want. This fact allows astronomers to crowd as much observation as possible into the precious few hours suitable for the purpose. It also frees astronomers for the time-consuming task of interpreting their observations.

Photographs have another valuable quality that is useful to astronomers. A plate can be exposed to the light from a star for as long as seeing conditions are suitable. The impression made on the plate increases in clarity and distinctness with the length of time of exposure. In this way, the photographic plate is able to "store" light, and this the eye cannot do. The eye is, however, more sensitive to momentary impressions and fine detail than is any photographic plate. The chemical grains of the plate are not nearly so fine as the retina of the eye. That is why observers see the canals of Mars quite easily, whereas photographs only rarely show this fascinating feature of the red planet.

Because of these and many other considerations, telescopes are made in a variety of designs and sizes. The optical or mechanical parts are especially adapted to the precise job the instrument is meant to perform. A large modern observatory includes a veritable arsenal of optical equipment of widely varying models and calibers. Studying the canals of Mars, for example, is best done through a visual refractor, whereas measurements of distant galaxies can only be accomplished through photographs made with a giant reflector.

These are the tools that modern science has developed to forge our bridge to the stars. The telescope is fundamentally the means with which we gather the light that emanates from distant heavenly bodies. This light is in turn magnified, focused, and otherwise made available to our eyes and measuring apparatus. Without the telescope, the stars would still remain as mysterious bodies floating through unexplained reaches of space.

These, then, are the tools fundamental to gathering and focusing the faint rays of light that come to us from the farthest reaches of space and time.

The heavens that these great artificial eyes have opened to us are far more magnificent and grandiose than that seen with the unaided eye. Galileo realized this the first time he peered at the heavens through his crude telescope and saw, for the first time, the bewildering riches that actually exist in the sky.

6

THE LANGUAGE OF LIGHT

Over the years, telescopes have become very sophisticated and powerful. The 200-inch Hale telescope, for example, can resolve the light filtering down to earth from galaxies that are as much as two billion light years away from us. This light, gathered and focused by the giant mirror, has been traveling through the vast expanses of space for two billion years. It began its journey when the first living matter was being formed in the lukewarm seas of the young earth.

We have already seen how telescopes gather light and bring the distant stars closer to man's eye. Now a single question remains: how are scientists able to learn so much about these distant stars through this single connection? How do scientists interpret the language of light?

To answer these questions, let us go back to Isaac Newton and his famous prism that produced the "celebrated phaenomena of colours." Actually, the answers are to be had by performing a simple experiment, one that

The Language of Light 61

you can easily do by yourself, with readily available equipment. All you will need is a sunny window, a large cardboard disk, and a glass prism—any triangular wedge of glass will do.

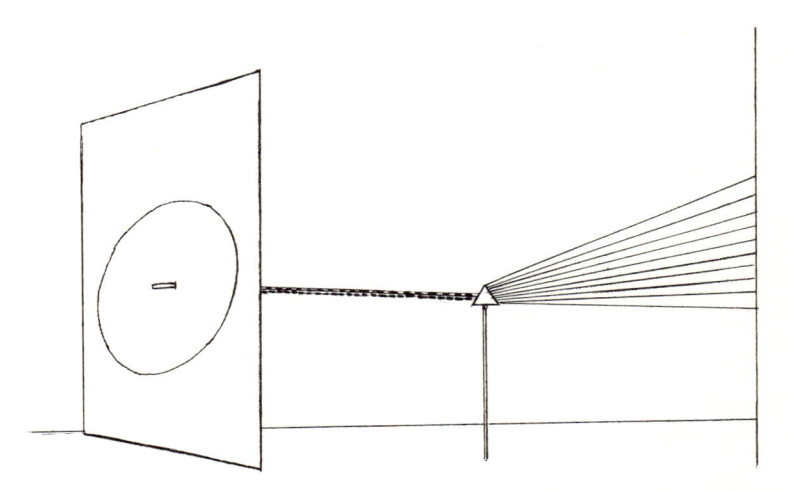

On a clear, bright day, when sunlight is streaming through your window, block out the light with the large cardboard disk. Then cut a narrow slit in the center of the disk. You will see a streak of light shine into the room through the slit and fall on the opposite wall in the form of a white line.

Set your prism—a triangular wedge of glass—in this streak of light. As soon as the prism is in place, you will see the white strip vanish. In its place there will appear a whole rectangle, below the spot where the strip of white light fell, made up of bands of brilliant colors. The top of the rectangle will be red, the bottom will be violet, and in the middle you will find a broad band of white. Now, narrow the slit in the cardboard through which the sunlight passes. As you narrow this slit, the white in the rectangle will shrink until it finally disappears. You will now have a rectangle made of bands of red, orange, yellow, green, blue, and violet which blend into one another without any sharply defined boundaries between them.

Scientists call this band of colored lights a "spectrum." When it is produced through the dispersion of ordinary white sunlight, it does not contain every color possible. Modern spectrographic techniques can, however, produce an unbelievable number of colors. The spectrum of white sunlight does, however, show all the simple colors from red to violet. These are the same colors that are produced in a slightly more blurred version by a rainbow, when sunlight is dispersed in the falling droplets of rain water.

To the naked eye, the spectrum appears to vanish at both ends of the colored rectangle. Above the red band and below the violet band, nothing remains to be seen. Actually, this illusion is due to a fault in our eyes. The nerve endings in our retina are sensitive only to this group of colors. Though our eyes cannot see beyond this range, other means of observation show that radiation continues beyond the red and violet limits of seeing.

An ordinary photographic plate, for example, is darkened when it is held below the violet edge of the visible spectrum. The chemicals in the plate react to what scientists call "ultraviolet light." If we pass the sunlight through a prism made of quartz crystal instead of glass, the area of ultraviolet darkening of the photographic plate can be extended.

Toward the other end of the spectrum, beyond the red band, invisible radiation can also be shown to exist. To prove this, we need a thermometer or, better still, a sensitive thermoelement. As we move this meter upward along the band of colors, we can measure the heat radiation striking the wall throughout the visible portion of the spectrum. Beyond the visible portion, heat radiation continues to strike the wall. The thermoelement is heated by what scientists call "infrared light." If we pass the sunlight through a prism made of rock salt instead of glass, the area of infrared heating can be extended. And, if we pass the light through a grating instead of a prism, the infrared range is extended even farther.

Long before we reach the limit of this infrared radiation, we can replace the thermoelement with an instrument that is sensitive to radio waves and show that the radiation of the spectrum extends even beyond the infrared radiation. This would, however, take us far beyond the limits of our experiment with a home-made apparatus.

In our experiment, we have discovered that ordinary sunlight can, with a little help, be broken down into a continuous sequence of radiations. Toward one end of this sequence, we see the violet band merging into the invisible ultraviolet band; toward the other end, we see the red band merging into invisible infrared radiation. You can easily see, now, how helpful it would be if we had some way of classifying these bands of radiation.

In the equations used to describe electrical fields, scientists found a system of classification that could be conveniently applied to light. Electric waves are distinguished from one another by wave length (the size of the oscillation of the wave) and by frequency (the number of crests that pass a given point every second). Electric waves are much longer than light waves, and they can be produced by special electrical apparatuses we call radio transmitters. These waves, whose lengths can range into miles, are used for radio communication and in radar. They are not perceptible as heat or light.

In the same way that we can measure the lengths and frequencies of electric waves, so can we also measure the wave lengths of light. This was first done by a Swedish physicist named Anders Jonas Ångström early in the nineteenth century. He studied the colored bands of light in the spectrum and, through an ingenious apparatus, was able to measure their wave lengths and frequencies. To this day, light waves are measured in what physicists call "angstrom" units—one angstrom is equal to 0.0000001 millimeter in length—in honor of the pioneer work done by this Swedish scientist.

Our eyes, Ångström discovered, are sensitive only to those wave lengths that fall between 4,000 and 7,000 angstroms. There is some variation in individual eyes which allows some people to see slightly more than others.

Beyond the red band, as we have seen in our experiment, radiation continues into what we call the infrared band. Infrared rays are longer than light rays and radiate heat instead of color. This heat can be measured when an infrared ray strikes an object. But, just as light is invisible until it is reflected from a surface, so are infrared rays cold until they strike an object. All things not at absolute zero temperature generate infrared rays.

These rays are very useful because they can easily pass through fog and other atmospheric obstructions. Photographers take advantage of this property and use film sensitive to these rays to take pictures at a distance where the haze of the earth's atmosphere becomes a problem or at night when there is no light. Today we have instruments

The Language of Light

so sensitive to infrared rays that scientists could detect an ordinary bathtub full of hot water on the moon.

Although infrared rays have longer wave lengths than visible light, scientists have discovered radiation with even longer waves. These are radio waves with lengths above 0.04 centimeters. We tune in these waves on television sets and short-wave radios. They are indispensable to our communication systems. Longer still are the waves in the band used for regular radio broadcasting. The longest waves of all are those that are broadcast by regular alternating electric currents. These can be measured in miles.

Infrared and radio waves are longer than visible light rays, but, as we have already seen, there are also shorter waves in the spectrum. These are the ultraviolet rays that blackened our photographic plate beyond the visible violet band. This experiment was first performed in England in the eighteenth century by William Hyde Wollaston. The darkening of silver chloride proved that there was radiation which had too short a wave length to be seen by the human eye.

These rays, which we now call ultraviolet, range between 0.00004 and 0.0000005 centimeters in length. You are probably familiar with the effect of ultraviolet rays from the sun. They cause a chemical change in the cell tissue of your skin. A little exposure means a handsome tan; too much means a bad sunburn. Ultraviolet rays also react with ergosterol, a chemical in the skin, to produce Vitamin D. Since this vitamin is necessary for strong bones, our bodies need the ultraviolet radiation of sunshine. Vitamin D is also formed when milk is exposed to ultraviolet light, and that is why many people drink irradiated milk.

Short-wave ultraviolet light can be stopped quite easily. The rays cannot pass through ordinary glass readily, and the atmosphere of the earth filters ultraviolet radiation. This is a good thing, because, though some ultraviolet light is necessary for our bodies, too much is dangerous. It can cause severe burns and will actually kill microorganisms.

This property of ultraviolet rays is useful for sterilizing rooms and equipment in hospitals and food-processing plants. If too much ultraviolet radiation were present in the world, it would kill the helpful microorganisms which are necessary for human life and thus disturb the balance that makes life possible.

Beyond the ultraviolet radiation, scientists have discovered even shorter rays. These are the X rays, which range in length from about 0.000005 centimeters to 0.00000001 centimeters. X rays have the amazing power of passing right through some solids. If you stand in the way of visible light, you will stop the rays, causing a shadow to form behind you. If you stand in the way of X rays, however, they will go through your flesh, though they will not be able to pass through your bones. The shadow of your skeleton is then recorded on film. Doctors use X rays when they want to look inside you to study your bones or internal processes. Too many X rays can, however, be very dangerous to humans, causing damage to cell tissue. That is why they are used only when absolutely necessary.

X rays were discovered by a German physicist named Wilhelm Roentgen while experimenting with a cathode tube. Though the practical application of X rays was quickly realized, it took many years of research before it was shown that these swift, penetrating waves were in any way related to light.

Beyond these X rays are what scientists call "alpha,"

"beta," and "gamma" rays. These rays are given off by such radioactive elements as radium, polonium, and uranium. They have far greater powers of penetration than X rays. Indeed, they can pass through such solids as metal and stone. Lord Ernest Rutherford, the English physicist, used these rays in the study of atoms. He passed them through every conceivable kind of material, thereby discovering the structure of the atom. Today, gamma rays are used by doctors in their fight against cancer. These

WILHELM KONRAD ROENTGEN (1845-1923)

rays are strong enough to kill living tissue, and, when they are carefully aimed at a cancerous tumor, they kill it as effectively as though it were cut out by surgery.

The most powerful radiation is provided by "cosmic rays"—derived from the Greek word *Cosmos,* meaning "the universe." These rays come from outer space and were first discovered by V. F. Hess, an Austrian physicist, in 1913. Though not too much is known about them, cosmic rays are important because they may give us additional clues to the formation of the universe. Scientists today are studying them in laboratories all over the world. Cosmic radiation is, of course, deadly. But, because they are absorbed by the earth's atmosphere, cosmic rays are not dangerous to man as long as he remains on earth. Above earth's atmosphere, however, cosmic radiation presents serious problems and remains one of the unknowns that must be solved before men can make extended voyages into space.

The entire range of radiation—from the tiny, extremely high frequency gamma rays, whose wave lengths are measured in millionths of a millimeter, to the long waves broadcast by ordinary alternating electric current, whose lengths are measured in miles—is called the electromagnetic spectrum. When all these types of radiation are plotted on a scale according to their wave lengths, we find a continuous sequence with no gaps between the various rays. One group blends into another, and it is impossible to find an exact dividing line. Red light waves blend into invisible infrared rays, infrared rays blend into radio waves, and so it goes throughout the spectrum.

Though these waves vary in length, frequency, and powers of penetration, they have many properties in common. All can be either refracted or reflected. None of them exist as they move through space, becoming discernible only after they are reflected from an object. All these waves travel at the same speed and react to gravitational or electrical fields in similar manners.

Visible light, as we can see from our complete electromagnetic spectrum, is only a tiny portion of the entire range of radiation. It occurs in a narrow band that is almost in the middle of the entire range. Though it consists of only a small portion of the spectrum, light is obviously closely related to the rest of electromagnetic radiation.

This fact spurred a great deal of scientific interest in light, since this is the easiest portion of electromagnetic phenomena to study. One of the scientists who was drawn to this problem was a Scotsman, James Clerk Maxwell, who lived from 1831 to 1879. Maxwell's early work was concerned with electricity; he noticed the similarity between electrical phenomena and light and developed a mathematical model that could be used to explain the work of Faraday and Oersted, who showed that, when an electric current flows through a wire coil, a magnetic field is produced around the coil.

Maxwell discovered that his mathematical model could also be used to explain the observed behavior of the movements of light waves. According to his model, a light wave is a series of alternating electromagnetic fields flowing through space. Jules Henri Poincaré, a French mathematician and physicist who elaborated on the equations of Maxwell, later showed that these fields change direction, or alternate, 1,000,000,000,000,000 times per second.

On the basis of the electromagnetic theory of radiation, scientists were able to predict the existence of radio waves before they were actually discovered. In 1887, Heinrich Rudolph Hertz, a German physicist who followed the lead of Maxwell and Poincaré, succeeded in producing such waves. At the turn of the century, the Italian scientist Guglielmo Marconi used them for sending wireless messages.

This, then, is the electromagnetic spectrum. It ranges from radiation whose wave lengths are infinitesimal to those whose lengths are measured in miles. Light rays represent only a small portion of this spectrum, but the language of light can tell us a great deal about its source.

7

BREAKING THE CODE

We know from everyday experience that only hot bodies give off light. A heated element in a light bulb, for example, glows white hot to produce light. We can feel the heat of a brightly shining flame just as we can feel the heat from the sun. There are some exceptions to this rule as, for example, the luminous dial of a watch. It glows in the dark even though it generates no heat. A firefly also glows in the dark with its own built-in cold light. The amount of light from this kind of source is, however, negligible. For the most part, we can say that the light and heat we use come from bodies at a high temperature.

To see how this phenomenon works, we need only watch an iron stove in a dark room. As long as there is no fire in the stove, it is a dead object that makes no impression on us aside from its general shape and size. Even when we light a fire inside the stove, it still makes no impression because we cannot see the flames. As the fire burns, however, heat gradually mounts, passing from the flames of the burning fuel to the iron lining. Soon the stove begins to show signs of life. We feel warmth strike our hands and face from its direction, and the sensation grows stronger and stronger, though we still cannot see the stove in the dark.

As the metal lining continues to heat, the stove gradually appears. At first it is only a pale, gray ghost. As we have already seen, where light impressions are faint, we can see no colors. As the fire continues to heat the stove, we see the outline of the lining in dark red. Gradually, the red color becomes brighter, and, if the fire in the stove is hot enough, we will see the color turn from red to orange and, finally, turn white hot.

Actually, we can perform this same experiment much more easily with a needle held in a gas flame. As the needle is heated, it passes through all the color stages more quickly than the stove, and, since we are not sensitive to the heat radiated by so small a body, we can safely watch the change from red to white heat. We can also produce the various colors in turn by shifting the needle from one part of the flame to another.

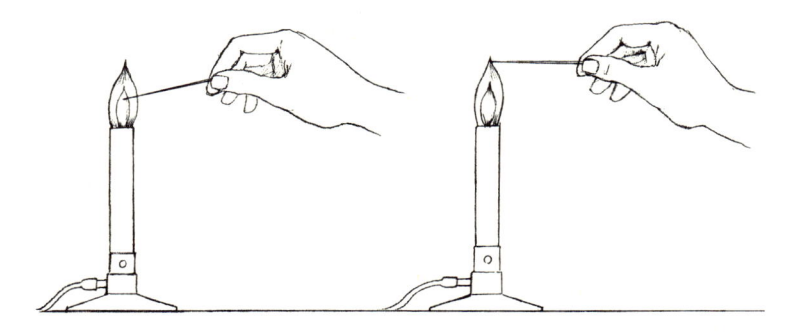

We see that the color produced by a heated body is directly related to its temperature. At the threshold of visibility, the body glows dark red and gradually brightens until it glows white hot as we raise its temperature.

The colors of the stars have exactly the same relationship to their temperatures. Stars, the astronomer tells us, are incandescent bodies with a huge range of size, temperature, and mass. They are not heated from the outside like our needle. Instead, they are more like our stove, whose hot interior radiates the heat outward.

It should not be difficult, then, to measure the temperature of a star. All we have to do is compare its color to that of an earthly body that is progressively heated. When we find the colors matching, we will know the temperature of the star. We should be able to do this except for one important fact: the temperatures of the stars are too high. Most earthly bodies will simply vaporize before they reach this temperature. All we can learn from such a comparison is that the stars are hotter than almost any temperature we can generate here on earth. We must look for another method in order to get a more precise temperature reading of the stars.

Astronomers have developed such a method. They can gauge the temperature of a star quite accurately. This method is based on carefully calculated laws of radiation that were worked out in laboratories, where temperatures can come nowhere near those of even average stars. These laws are based on the fact that the temperature of a radiating body determines which segments of the electromagnetic spectrum it radiates. The hotter a body becomes, the shorter the wave lengths it broadcasts.

We can show that this is true by performing an experiment with comparatively simple apparatus. We need a prism, a darkened chamber with a slit to admit light, and a suitable surface on which to project the spectrum. For our light source, we attach a thin wire connected by heavier wires to a source of electricity. One of the thick wires is run through a variable resistor so that we may regulate the strength of the electric current. Finally, we need an ammeter to measure the current.

We begin our experiment with a high resistance. The current running through our wires is weak, and the wire does not even grow warm. We gradually reduce the resistance and wait for something to appear on the surface where we project the spectrum. We know that, as more current flows and the temperature of the wire rises, radiation will first show itself in the form of heat. To measure the appearance of heat radiation, we place a thermoelement close to the screen, above the level where we expect the visible spectrum to appear.

Soon, the thermoelement will register the appearance of heat radiation. As still more current flows through the wire, we suddenly see something; a red strip appears on the screen—a red image of the filament which can now be seen to glow. The red strip grows brighter as the filament grows hotter. It gradually broadens downward, and the descending edge of the band changes to orange and then yellow. As the spectrum continues to broaden, green is added and finally—always by way of in-between shades that fade into one another—blue and violet. By this time, the wire filament is white hot.

Meanwhile, as the spectrum broadens, the heat radiation registered by the thermoelement also mounts. The total amount of heat radiated increases with the temperature, and, at the same time, shorter and shorter waves are added to the initial long ones. If we can make our filament hot enough, the shorter waves will predominate and the greatest intensity of radiation will move into the blue or violet.

Although the relation between radiation and temperature is not difficult to demonstrate, it is far from being simple. It took an exhaustive examination of this radiation phenomenon before scientists arrived at the formula by which they derive a temperature value from the measured intensity of radiation. Every temperature, according to this formula, has its characteristic curve when plotted on a graph. This curve shows the relationship between temperature and intensity of radiation. Scientists have plotted these curves for practically every temperature value from absolute zero to that which prevails in the hottest stars.

Now we have the means by which we can take the temperature of a star. All we need is the spectrum from the light of a particular star, which can be obtained from the light gathered by telescopes. Then, to determine its temperature, all we do is measure the intensity of the radiation at various points on the spectrum and determine at which wave lengths the radiation is strongest. Finally, we search through our collection of curves to find the one that corresponds to the observed wave length, and we have the temperature.

The star temperatures at which we arrive by these methods are, as we expected, extremely high. The coldest —and therefore the reddest—stars have about the same temperature as the incandescent filament in an electric light bulb (about 3600 degrees Fahrenheit), whereas the hottest white stars show temperatures that exceed 55,000 degrees Fahrenheit. Because of these very high temperatures, we cannot compare the spectra of most stars to the spectrum of some earthly source of radiation. Consequently, we have to depend on the accuracy of our mathematics in this indirect method of measuring their temperatures.

Once we know how hot a star is, we can compute its size and distance. Both values are determined by comparison with the sun. We know how large the sun is, and we know how much light it radiates. By comparing these figures to those of a more distant star, we can estimate its size and its distance. In this way, astronomers have been able to gain a general idea of the sizes prevailing among the myriad stars of the heavens. Our sun, it turns out, is among the smallest of stars, though it is one of the commonest single types.

We have seen how important the spectrum is in determining the size and distance of a star. When we try to learn what a star is made of, the spectrum becomes even more important. In dealing with the problem of a star's composition, we begin to realize how rich the language of light actually is.

Thus far, we have discussed only the continuous spectrum of light from the stars. When light contains all the wave lengths of the visible part of the spectrum, we produce this continuous band of light. The spectrum of most stars is, however, not continuous. There are gaps, mysterious black lines, throughout the spectrum. Some of these bands are quite broad, but most are narrow. What can these black lines mean?

To answer this question, we must go back to our first experiments with the prism. Newton's "phaenomena of colours" fascinated many scientists who tried countless variations of the original experiments. One of these variations involved heating a single element or compound until it glowed and then passing this glow through the prism.

The element sodium, for example, was found to have a characteristic yellow line at the place where the brightest yellow would appear in the continuous spectrum. This yellow band, when magnified, was found to be divided. From this, scientists concluded that sodium radiates only in a single wave length that produces this particular band of color in the spectrum. The sodium atom that produces this band can only vibrate in this characteristic rhythm.

The same was found to hold true for all the other elements when they were heated to incandescence. Their spectra consist of single lines of the continuous spectrum. Some atoms show a few of these lines; some show a great many. The spectrum of iron, for example, reveals almost a thousand of these lines.

In this way, all the elements have been subjected to spectrum analysis. Each element has its own characteristic series of wave lengths. For example, only gold "broadcasts" at 2427.95 angstroms and 2675.95 angstroms, whereas only sodium has a wave length of 5889.95 angstroms. A few of the elements have wave lengths that are very close to one another. One of the wave lengths of silver is 2246.412 angstroms, which is almost exactly the same as one of the wave lengths of copper, which is 2246.995 angstroms.

Out of this research came the spectroscope, an instrument that allows us to see the spectrum of a particular element or compound more clearly. It consists of an arrangement of lenses that magnifies the pattern of the spectrum and a scale with which the position of each line can be measured. The spectroscope is used to discover exactly which elements are to be found in any compound, for each element broadcasts electromagnetic waves only of characteristic length. When we see a yellow line, we may be sure that sodium is present. A red line and a broad violet line are characteristic of potassium.

Although the spectroscope is an important instrument in chemical analysis and is widely used in scientific crime detection, its most important use is in the analysis of stars. It offers us a clue to the meaning of the mysterious black lines that appear in the spectrum of the sun and the other stars.

These lines were first noticed by the German physicist and optician Joseph von Fraunhofer who, while using an early form of the spectroscope, noted their appearance in the spectrum of the sun. He was able to count about 750 parallel lines. Fraunofer did not know what to make of them, and it was fifty years before the cause of these "Fraunhofer lines" could be explained.

Another German physicist, Gustav Robert Kirchhoff, discovered the answer. He heated an element until it produced a colored light. Then he passed a beam of this light through a vapor cloud of the same element. When the beam continued through a prism after passing through the cloud, it produced black lines in the spectrum where the characteristic color bands of that element would normally appear. Kirchhoff concluded from this observation that, when light from a given element passes through a cloud of vapor from the same element, it is absorbed in the process. The black lines that appear in the spectrum show where the characteristic bands should have been.

Here, then, was a means by which scientists could analyze the chemical composition of the sun. When sunlight passes through the atmosphere surrounding the sun, Kirchhoff reasoned, a similar thing must happen. Elements present on the sun and in its atmosphere appear in the

※ *Breaking the Code* ※ 79

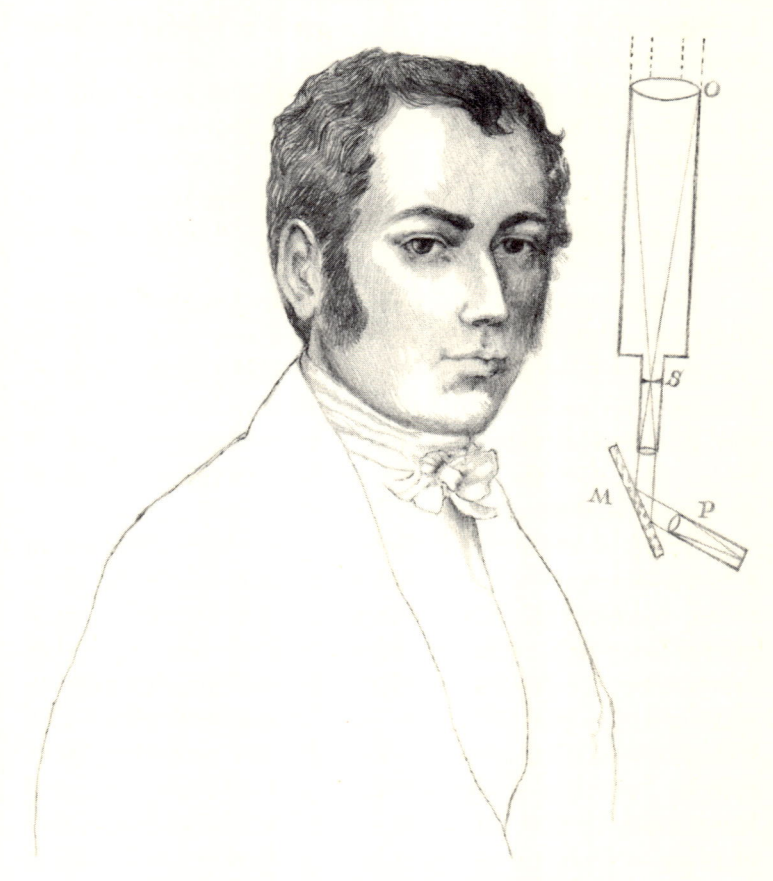

JOSEPH VON FRAUNHOFER (1787-1826)

spectrum as Fraunhofer lines instead of as colors because of the absorption that takes place.

Using a modern spectroheliograph, astronomers are able to pick out about 60,000 Fraunhofer lines in a distended spectrum of the sun. Because there are so many, it is almost impossible to pick out those that belong together, thus recognizing the elements that they represent. The single element iron, for example, has almost one thousand characteristic lines. In order to make sense out of all the Fraunhofer lines present in sunlight, scientists must painstakingly determine the wave lengths of the lines that are characteristic of each element and then find their corresponding position in the sun spectrum.

In this way, the lines of some fifty elements have already been detected in the sun. We are sure that these elements are present in the sun or its atmosphere. We cannot, however, be sure that other elements do not exist in the sun. An element may be lacking in the atmosphere and yet be present inside the sun, or it may be present in such small quantities that its absorption leaves no visible trace.

Some of the missing elements have lines that fall in a part of the sun spectrum that we cannot see—in the ultraviolet, for example, that is absorbed by the earth's atmosphere. Since this is the case for many of the elements that are missing, we can assume that nearly all the elements on earth are also present in the sun. Their proportions also seem to be similar to those on earth. An exception, however, is hydrogen, a part of water and of many chemical compounds in the plant and animal worlds. Though this light gas is plentiful on earth, it is altogether predominant in the sun and its atmosphere—making up more than 50 per cent of the mass.

By analyzing the spectrum of the sun, scientists can determine its chemical make-up. Exactly the same method is used to analyze the content of the more distant stars. As long as our telescopes can resolve enough light to form a spectrum, we can get a fairly accurate idea of what the source of that light is composed of, no matter how far away it may be.

Though the language of light has proved to be far more rewarding than early astronomers imagined it to be, it still represents only one window through which we look out on the universe. There is, however, a second window that science has revealed from which we also look to the stars.

8

WINDOW BEYOND SIGHT

Light represents only a tiny portion of the entire electromagnetic spectrum. But it is the portion of the spectrum that man has studied longest and knows most about. By interpreting the language of light, astronomers have been able to learn a good deal about the universe around us. They analyze the chemical content of stars that are millions of light years away from us and, through visual observation alone, estimate their sizes and masses.

Visible light, however, is only one of the windows that looks out on the universe. Within the past few decades, scientists have developed the instruments and techniques that have opened a second window. This window falls in that part of the electromagnetic spectrum that we recognize as radio waves. Today, giant radiotelescopes with huge antennae gather these waves that are broadcast by even the most distant stars and galaxies and make them available for analysis and interpretation. Radio waves are now being used along with light waves to build a broader bridge to the stars.

It is only an accident of physiology that makes our eyes sensitive to light waves. They might just as easily have been sensitive to radio waves. If this were the case, the heavens would appear very different to us. In fact, they would be almost unrecognizable. The sun would be much dimmer than it appears to our light-sensitive eyes, whereas the Milky Way, that nebulous band of light that we can barely see on a moonless night, would become like a brilliant neon sign blazing across the sky. It would be so bright that the position of the Milky Way relative to the rotation of the earth, rather than our position relative to the sun, would determine night and day.

Hundreds of now-invisible stars would populate the heavens, forming completely unfamiliar constellations. We would be able to see the great turbulent clouds of interstellar gas and the billowing remnants of exploded stars. One of these, the crab nebulae, for example, would provide a brilliant celestial display easily visible to our radio-sensitive eye, instead of the wispy cloud that can be seen only with the aid of a powerful telescope. A dazzling object in our radio sky would be a pair of galaxies in collision some 700 million light years away. Visually, these colliding galaxies are so faint that they can be photographed only by long exposure under the largest optical telescopes. The radio energy that is broadcast by this source (called Cygnus A) is so immense that it surprises even the astronomers, who are accustomed to dealing with astronomical figures.

In a single second, Cygnus A emits an amount of radio energy which, if concentrated and translated into heat, could supply all the earth's power requirements for the next trillion years. Its radio emission is so strong that it could easily be detected by a radiotelescope even if it were fifteen times farther away than it is—a distance far beyond the range of our largest optical telescope.

Cygnus A provides a dramatic demonstration of the new window that the radiotelescope provides. Radio "sight" opens to astronomers a greater range of space and distances than can be explored visually. It also brings into view for the first time many close celestial bodies that are too cool to be seen visually. Because radio waves are so much broader than light waves, they can pass through the interstellar clouds that block our vision in large portions of the universe. The center of the Milky Way galaxy, for example, is blocked from view by the huge cloud in the constellation Sagittarius. Radio waves, however, have penetrated this cloud and provided astronomers with their first glimpse at the heart of our galaxy.

Radioastronomy is still a young science. Today, it is in approximately the same stage of development that visual astronomy was in the years after Galileo invented the first astronomical telescope. It has opened a new window, but this window is still hazy and unclear. One of the big handicaps of the radiotelescope is its inability to locate or resolve objects precisely. Our present instruments cannot pinpoint radio stars or other sources of radiation; their sharpest resolution can only define the source diffusely; however, radiotelescopes are developing swiftly. Larger and more ingenious instruments are being built all the time, and we can expect improvements in resolving power that may soon approach that of visual telescopes.

The entire field of radioastronomy began innocently enough when, in the spring of 1935, Karl G. Jansky, an engineer working at the Bell Telephone Laboratories in New

Window beyond Sight 85

York, was making a series of routine observations of the interference experienced by a radio communications link. Jansky discovered that the interference was being caused by radio waves of a characteristic wave length that seemed to be coming from the direction of the Milky Way, or the central plane of our galaxy.

KARL G. JANSKY (1905-1950)

Fascinated by this mysterious phenomenon, Jansky made a study of these waves and discovered that they were indeed coming from regions of space outside the solar system. Although his proof of this theory was decisive, he could not find out much more about these radio waves, and his discovery was, unfortunately, overlooked. The radio-receiving techniques of that time were simply inadequate to interpret these strange waves that filtered down to interfere with radio communication on earth from the far reaches of our galaxy.

Jansky's discovery remained almost unknown, a bizarre natural phenomenon without any practical value to man. This unexpected gift of nature to man lay disregarded by the world's astronomers. Then, in the course of World War II, the development of radar provided receiving equipment of a sensitivity and excellence surpassing anything that was previously available. As soon as these sensitive new instruments were used to study the radio waves from the sky, it was quickly realized that Jansky's accidental discovery had opened an entirely new avenue for the exploration of space.

The improved definition and sensitivity of this equipment enabled astronomers to make an entirely new chart of the heavens. They discovered that the concentrations of radio emission in particular parts of the sky did not always correspond with visible phenomena. Bright stars such as Sirius or Capella were found to be poor sources of radio waves, whereas other celestial bodies which cannot be seen at all were discovered to be powerful broadcasters.

Some of these invisible radio wave sources have been identified as dark stars. Others, like the huge cloud in Cassiopeia that was first believed to be a "radio star," were shown to be "irregular clouds of gas with violent internal motions and high excitations." The dark stars discovered by radioastronomy are so cool that they give off no visible light, or not enough to be seen even through our most powerful telescopes. The difficulty in relating these celestial objects to the universe as perceived visually arises partly because the objects that give off radio waves are generally very dim visually and partly because radio waves may reach us from distances which are beyond the range of optical telescopes. No clear-cut method has yet been developed to measure the distances of a particular radio source.

Another source of radio waves was discovered by a young Dutch student named Van de Hulst, during the terrible years of the German occupation of his country. He made a series of calculations which led him to predict the existence of radio waves given off by neutral hydrogen atoms. He showed that, though the emission of waves by an individual atom is very rare—occurring once for each atom in every 11,000,000 years—the number of atoms present in interstellar space is so great that the total emission should be detectable. Years of peace were required to develop the right equipment, and then, in 1951, these weak radio signals given off by hydrogen atoms were detected—an event which represented a triumph of technical skill and a brilliant proof of Van de Hulst's calculations.

Dutch astronomers at the observatory in Leiden, under their director Jan Oort, seized on this particular wave length broadcast by hydrogen atoms to map the structure of our Milky Way galaxy. The large optical telescopes are powerless to penetrate the interstellar dust clouds which obscure the structural details of the system, and visual exploration is further handicapped by the underprivileged position of our solar system on the edge of the galactic disk.

The interstellar dust presents no obstacle to the twenty-one-centimeter waves of hydrogen which can pass right through them the way an X ray passes through flesh. Using this window, Oort and his staff have succeeded in mapping our galaxy with almost unbelievable detail and elegance, describing, in the process, parts of the Milky Way which man will never be able to see.

Radiotelescopes also pick up signals that are broadcast from sources far beyond the confines of our Milky Way galaxy. One of the strongest of these extragalactic radio sources is the emanation from the colliding galaxies in Cygnus A which we mentioned earlier. This source was carefully measured and located by Prof. Martin Ryle and his colleagues in Cambridge with a huge radiotelescope. Their measurements were accurate enough for the astronomers to use the 200-inch telescope on Mount Palomar to make an intensive search of this region of the heavens.

The 200-inch Hale telescope then discovered a remarkable celestial event. Two great extragalactic nebulae seem to have collided with each other. This collision, which is at a distance of some 700 million light years, is close to the limit of clear definition of the world's largest telescope. Yet, the radio waves that come to us from this source are so strong that they could be detected even if the source were fifteen times as distant. This fact leads astronomers to hope that, with radiotelescopes, they will be able to observe phenomena that are far beyond the range of optical instruments.

The range of the Hale telescope represents our limit of visual contact with the universe around us. The atmosphere of the earth together with several other factors prevents optical telescopes from seeing clearly beyond this limit regardless of how much larger we make them. When telescopes are eventually mounted in orbiting satellites or an observatory is set up on the moon, resolution of distant celestial bodies will be greatly improved. It is doubtful, however, whether even these instruments will be able to see farther than the Mount Palomar telescope does.

Radiotelescopes, however, do not have such limitations. Indeed, Prof. Bernard Lovell, director of the Jodrell Bank Radio Observatory in England, says: "Some recent measurements of unidentified radio sources imply that we are dealing with events at distances of many thousands of millions of light years. Indeed, we may now be in the process of probing the ultimate depths of space and time."

The 250-foot radiotelescope at Jodrell Bank is now the largest operating instrument of its kind in the world, but it is not destined to maintain this distinction for long. A giant 600-foot instrument for the United States Navy is now nearing completion in a West Virginia valley.

This radiotelescope will be the largest steerable, paraboloid dish antenna ever built. Standing nearly 700 feet high, its reflector will be 600 feet in diameter, or more than seven acres in area. The installation of this instrument is expected to cost more than $100,000,000. The structure will require 20,000 tons of steel, 600 tons of aluminum, and 14,000 cubic yards of concrete. The reflecting dish will be cradled in two structures resembling giant ferris wheels which will tilt it to any angle of elevation from 0 to 90 degrees.

The entire structure will ride on four trucks whose wheels will roll along a circular railroad track nearly a third of a mile long. This track will enable the huge instrument to turn through a full 360 degrees so that the reflector may point in any direction. An aluminum wire screen will provide the reflecting surface of the paraboloid. The screen will be divided into panels whose positions will be automatically adjusted by servomechanisms to compensate for any distortions caused by wind, gravity, or temperature changes. In this way, the surface will be maintained as a paraboloid to within three-fourths of an inch, which is the accuracy needed to work with twenty-one centimeter radio waves.

Other types of radiotelescopes are also under construction at various sites. In Aracaibo, Puerto Rico, for example, a 1,000-foot spherical dish antenna is being scooped out of the side of a hill. On completion, this radiotelescope will be used for radar studies of the planets. Of course, it will not be so versatile as the West Virginia instrument, because the dish will not be steerable. It will be able to scan only those parts of the heavens that it points to as the earth rotates on its axis. Its sheer size, however, will make up for its lack of maneuverability, and it is expected to enrich our knowledge of planetary temperatures, rates of rotation, and surface natures.

Radioastronomy is little more than fifteen years old, but it has already provided an invaluable store of information about the stars and the universe around us. By opening a new window in the electromagnetic spectrum, the radio telescope has provided a new link to the stars. Our bridge has become broader, and, with this new tool, man's vision is extended far beyond the current range of optical telescopes, allowing him an even more profound glimpse into the operation of the universe.

So far in this book, we have dealt only with light as the bridge to the stars. Light is, however, even more versatile. Not only does it provide us with a window on the universe—on the macrocosm of huge values of space and distance—but it also opens a window on the atom—on the microcosm where time and space shrink to nothing. Just as we could never begin to understand the universe of stars and galaxies without light, so we could not begin to understand the infinitesimal world of the atom without this bridge, for light led science from the galaxies into the heart of the atom.

9

COHERENT LIGHT, THE NEW TOOL

Light, then, is a quite remarkable thing. It enables us to see; it carries a code by which we can read the chemical makeup of a star so far away that it takes a million years for the light from that star to reach us; and it gives us a wedge with which we can pry open the innermost secrets of the atom.

For all this, however, ordinary light is a wild and untamed segment of the electromagnetic spectrum. The trouble lies in the fact that ordinary light is a chaotic bundle of waves. These waves come in an uncontrolled mixture of lengths, frequencies, colors, and directions. They reinforce or cancel one another at random, without rhyme or reason. Because of this undisciplined quality, light cannot be put to useful work at a given frequency the way, for example, a radio wave can be used.

Theoretically, however, a light wave is an ideal medium for communication. Light waves are tens of thousands of times shorter than radio waves. Indeed, a narrow band

Coherent Light, the New Tool

of light waves can hold trillions of cycles per second and thus should be able to "broadcast" enormous quantities of information. It has been estimated that a single beam of light—of tamed light, not ordinary light—could carry as many messages as all communication channels in existence today.

But, even more important, a light beam is also an ideal carrier of energy. Theoretically, a tamed beam of light should be able to transmit huge amounts of energy the way a radio beam transmits electron impulses. You can see an example of this ability of light when you focus an ordinary beam of sunlight through a magnifying glass. The focal point becomes hot enough to start a fire. Sun furnaces, using giant reflectors, achieve some of the highest temperatures man has been able to generate under controlled conditions over a lengthy stretch of time. Tamed light, not wasting its energy in random emission, should be able to transmit energy much more efficiently.

Despite all this potential, light has thus far never been harnessed for useful work. We have studied light, analyzed it, manipulated it, but we have never really gained any control over its chaotic waves. This question of control is basic to utilizing anything for useful work. Radio waves, for example, can be controlled. We can generate radio waves in a coherent manner. We can broadcast a beam having a particular frequency and parallel direction and use this beam to transmit messages; we use high-frequency radio waves to transmit television images; and we can use even shorter wave lengths in radar. Until we learned how to generate coherent light, we could never do the same with light waves. Today, however, this miracle has been attained. Scientists have tamed light and are learning how to apply light's fantastic frequencies to man's needs. This development has come about through the invention of the laser ("light amplification by stimulated emission of radiation"). This invention is as important a milestone in science as the invention of the vacuum tube and promises as many exciting developments as we have gained in electronics through the application of the vacuum tube.

With the laser scientists can, for the first time, generate coherent light. This light can be tailor-made in such a way that of its frequencies are exactly alike and precisely in step with one another—crest to crest and trough to trough, flowing in parallel waves of unequaled discipline.

Unlike so many important scientific breakthroughs of the past, the development of the laser was no accident. Everything about the laser represents a triumph of scientific theory and application. By logical deduction and the ingenious use of physical principles known for more than thirty years, Dr. Charles H. Townes, then at Columbia University, and his brother-in-law, Dr. Arthur L. Schawlow, of the Bell Laboratories, worked out the essentials of the laser in 1958.

They showed exactly how a coherent-light beam could be generated. After that, it was only a matter of time before a working laser was produced—in this case, two years. In 1960, Dr. Theodore H. Maiman, of Hughes Aircraft, produced the first coherent-light beam in history.

To understand exactly how this was done, we have to go back to the beginning of the twentieth century when the entire edifice of physics was being rebuilt. It all began with a German physicist named Max Planck. He was studying black light and discovered that, instead of flowing in a continuous stream as classical physics predicted it should, it flowed rather in tiny, discrete "bundles." Planck called these bundles "quanta."

*   *Coherent Light, the New Tool*   *   95

Quantum theory developed out of this important discovery. A quantum, Planck theorized, was the smallest unit of energy that exists in nature. This unit, according to the scientist, was basic and indivisible. All energy levels could be expressed in units of quanta, but never in fractions of quanta. Energy could have any number of quanta, but never a quarter of a quantum, never a fraction.

Planck's ideas were taken up a few years later by Albert Einstein in his explanation of the photoelectric effect. Einstein showed, through his mathematical model, that ordi-

NIELS BOHR (1885-1962)

nary light was also divided into separate, discrete units that he called "photons." All light waves, according to this theory, can be expressed in terms of these tiny "bundles" of light energy, but never in fractions of photons.

Later, Niels Bohr, the Danish physicist, applied these same ideas to the atom. The atom, according to Bohr's view, consisted of a heavy, sun-like nucleus with a cloud of electrons revolving around it in certain prescribed orbits. An electron of a particular atom may have any number of possible orbital paths around the nucleus and move from one level to another as it either gains or loses energy. In both cases, however, the electron could leap only from one level to another; it could never settle in an orbit in between. This model of the atom proved a boon to science. Out of it came the atomic bomb, with its inexhaustible store of energy, on the one hand, and the microscopic transistor, with all of its exciting applications, on the other.

Light, it was soon discovered, was closely tied up with these jumping electrons. In an ordinary atom, an atom that is quiescent, there is no light emitted. If, however, the electrons in orbit about the atomic nucleus gain energy—no matter how—they jump to a higher level. At this new energy level, the orbit is not stable. Within a few microseconds, the electron falls back to its "normal" orbit, giving up the "borrowed" energy in the form of electromagnetic energy. When this radiation falls within the visible portion of the spectrum, we see it as light.

To produce light, then, electrons must first absorb energy from an external source. That source may be electricity, as when you switch on a light bulb, or it can be light, as when you shine a bright light on phosphorescent material. In this case, light energy excites the electrons, and the material glows as the electrons fall back to their previous levels. Or energy can come in the form of heat, as when we make a lump of iron glow red hot in a furnace.

In all these cases, however, the light that results is incoherent. It is broadcast in every direction and at many frequencies. We are still faced with the problem of taming this chaotic stream of radiation. The solution to this problem lies in the energy levels of the atomic electrons.

We know from quantum mechanics that every atom has only a given number of energy levels for its electrons. Some atoms may have many such levels, whereas others have only a few. Each atom, however, broadcasts on a specific wave length or, if its electrons have many levels, on many specific wave lengths.

Here, in this physical principle, we have the first clue to the production of coherent light. If we can stimulate a certain atom to emit light at a specific energy level, and if we can do this to a number of such identical atoms at the same time, we will have light of a given frequency. This is not difficult to do. We can generate this kind of light with a phosphorescent material. When this material glows, electrons are falling back to a previous level, and, since this is happening with only one element, most of the resulting light is of the same frequency.

Now we come to the difficult part. We have a light source of a single frequency. This light is, however, being given off in all directions and at random as each electron falls back to the previous level in its own good time. If we could only get these frequencies in step with one another and parallel, we would have our coherent-light beam.

To get these waves into step with one another called for an ingenious technique. Light waves, theorized Townes and Schawlow, are subject to resonance in exactly the same way that sound waves are. When you play a note on the piano, for example, a tuning fork in the same room tuned to that note will vibrate in sympathy with the note sounded on the piano. This is an example of resonance. The tuning fork is stimulated to vibration by sound waves of its given frequency. The same thing can be done with electrons. When an electron is in a high-energy state, it can be stimulated to give up its excess energy in resonance to an identical frequency. An electron at a high-energy state is unstable and ready to give up its excess energy. It will do this automatically if left alone. If, however, we expose these excited electrons to electromagnetic waves of the same frequency as those they are waiting to emit, the electrons can be made to release their excess energy in a controlled fashion. Even more important, the radiation triggered by this kind of resonance will always fall exactly into step and reinforce the triggering waves.

In Dr. Maiman's first laser, a thin rod of synthetic ruby wound about with an ordinary photographer's flash tube was used. The tube's light was the source of energy used to excite the electrons in the ruby rod. Synthetic ruby is made, in part, of chromium. When the light source was flashed on, the electrons in the chromium atoms were pushed to higher levels of energy. As the electrons fell back to normal levels, they emitted light—in this case, a frequency that produced a visible red glow. This is an example of natural fluorescence. Although it had the same frequency for the most part, the red light emitted by the ruby rod was as incoherent as any other type of light. It shot out in every conceivable direction.

In order to tame these unruly light waves, Townes and Schawlow worked out a deceptively simple system. In the ruby rod, they reasoned, they had a source of waves of identical frequency. Some of these waves were bound to be coherent—that is, in step with one another—but they would soon be lost in the general disorder of incoherent light. If, however, some of these waves could be fed back into the ruby and made to go back again and again by placing mirrors at the ends of the rod, a potent chain reaction could be triggered.

The waves shooting off at odd angles would be lost, but those that were in the path of the reflected beams would soon build up. They would stimulate more and more electrons as they passed back and forth between the mirrors, stimulating the emission of light waves that fell exactly into step. In this way, the light waves could be amplified billions of times. If one of the mirrors, the scientists decided, were less heavily silvered than the other, the light waves built up by the reflected beams would finally burst through the mirror in a powerful beam of highly coherent light.

This was exactly what happened in a fraction of a second when Dr. Maiman turned on the switch of the laser he devised in 1960. The ends of the ruby rod were mirrored, with one of the mirrors made partly transparent. When the light in the flash tube went on, a brilliant red beam shot out of the lightly mirrored end for half a thousandth of a second.

That was the beginning. Dr. Maiman's device provided the proof for the theoretical model Townes and Schawlow had worked out. Coherent light *could* be made on order. Dr. Maiman's device was, of course, a prototype —a crude beginning. His laser, for example, generated so much heat and power that it had to be operated at liquid-nitrogen temperature, 196 degrees below zero Centigrade.

Within six months of the first laser, improvements

on the original appeared. Dr. Ali Javan of the Bell Laboratories produced a gas laser which could operate continuously, instead of in intermittent pulses as the first laser did. Dr. Javan's device looked something like a neon tube with mirrors at both ends. It used a mixture of helium and neon gases, activated by an electric current instead of light, and produced infrared light.

Dr. Javan's laser was soon followed by others based on other gases which emit light at varying frequencies. Today, work on the construction and design in lasers is proceeding in laboratories all over the world. These devices are getting smaller and lighter. Indeed, a laser developed simultaneously by scientists at IBM and General Electric is so small that it can hardly be seen without a strong magnifying glass.

With the development of the laser, light has finally been harnessed for work. A beam of coherent light can mobilize an enormous amount of heat when focused on a small spot. A beam from a ruby laser, for example, can be even hotter than the sun's surface. This power was dramatically demonstrated at the General Electric laboratory when a beam of coherent light burned tiny holes in diamonds. The red flashes lasted only a tiny fraction of a second, but it was long enough to have the target go up in a puff of blue smoke.

This quality has also proved valuable in medicine. Eye specialists at Columbia-Presbyterian Medical Center have used laser beams in delicate operations. The searing heat has been used to destroy tiny tumors on a patient's retina instantly and painlessly. The light went right through the eye's lens without harming it and then focused exactly on the tumors in the back of the eye.

In experiments on animals, doctors have used laser

beams to weld a detached retina to the back of the eye. A beam of light can be controlled with almost uncanny precision, and, because of this fact, coherent light may replace the scalpel for certain kinds of very delicate surgery.

Laser light has also been put to work in industry. The miniaturization of equipment, especially for the space program, has posed an almost insurmountable problem to manufacturers. Now, in this new kind of light, they have discovered a solution to many of their problems. A laser beam can be used to cut almost invisible parts of tiny electrical circuits; it can weld microscopic wires; and it can be used to drill holes with an accuracy never before possible.

A laser beam has been shot to the moon. This was done by engineers at the Massachussetts Institute of Technology in the summer of 1962. The experiment proved that coherent light can be focused with extraordinary precision by the use of simple lenses. The beam lit up a spot only two miles in diameter on the moon's surface. In contrast, the tightest microwave beam in our most advanced radars would have spread over an area more than 500 miles wide during the transmission across this same distance.

In chemistry, laser promises to open entire new fields of research. Laser light could conceivably produce completely new chemical reactions by heating only one element in a chemical mixture. It may be used to generate the heat necessary to spark controlled thermonuclear reactions, and it is already the most accurate measuring device known to man.

Coherent light is a new development. Its potential, as we have seen, is almost limitless. The laser has given man a new level in his control of light, and this new order of control will open up uses for light that we cannot even imagine at this time.

10

IN WONDERLAND

By the end of the nineteenth century, scientists believed that they had solved all the physical problems of the universe. The laws that governed the operation of this universe had been spelled out in elegant detail. They worked. That is, they corresponded to observation right down to the third, fourth, even fifth decimal point—which was certainly accurate enough.

The mystery of light was also solved for all time. In 1865, James Clerk Maxwell showed the world that light is an electromagnetic wave. His equations proved that a change in electromagnetic force generates waves that travel through space. He showed that light was just one portion of the electromagnetic spectrum, corresponding to a characteristic band of frequencies.

Maxwell's formulas accurately predicted how light, heat, and all other electromagnetic waves travel through space. They also predicted the existence of many other frequencies in the spectrum, including radio waves and the broad bands broadcast by alternating electric current. These waves differ sharply in the way they are generated and detected. We can see light with our eyes, feel heat with our bodies, and receive radio waves with electronic devices. These differences are, however, only superficial. Maxwell proved that they all represent electromagnetic energy, traveling through space as waves.

JAMES CLERK MAXWELL (1831-1879)

The only real differences in the parts of the electromagnetic spectrum lie in wave lengths. These lengths vary, however, over a huge range. They include mile-wide waves broadcast by alternating current and X-ray waves that measure less than a hundred of a millionth of an inch. All these waves, according to Maxwell's equations, range continuously across the spectrum with no sharp divisions between them. The shortest radio waves, for example, are the same as the longest heat waves.

With wave theory, scientists could explain most of the observed phenomena of light. Diffraction, for example, was accounted for by this theory. We know that light does not cast a precise shadow. If you examine the edge of a shadow, you will see that the change from light to dark is gradual, one merging into another. This is exactly what is predicted by Maxwell's equations. They show that waves bend as they pass an obstacle, never stopping abruptly, but fading gradually. You can see this effect in the ripples that form in a pool of water when you drop a stone into the pool. The waves spread out in a circle. When you place an obstruction in the path of this advancing wave, it becomes "shadowed" by the obstruction.

Wave theory opened a window on the universe. With this powerful theoretical tool, scientists were able to measure and analyze the chemical content of even the most distant stars whose light could be resolved by a telescope. Each wave length of light was shown to be the herald of a particular element, and everything, scientists believed, could be learned simply by refining and perfecting the techniques that were already at hand. Newton's old concept of light as tiny corpuscles was all but discredited. Only few scientists gave it any serious thought at all, and then only as an interesting historical oddity.

The edifice of science built with such painstaking labor and thought seemed to be unassailable. The mechanics of the universe had been reduced to provable formulas. All that remained to be done, the learned professors gloated, was to tidy up a bit in the corners—to repeat an experiment here, straighten out a discrepancy there.

Many of these discrepancies involved light. Wave theory, for example, could not satisfactorily explain such an apparently simple thing as a hot object. It was known that heat makes a thing glow and that the color of this fiery light is determined by the temperature. Wave theory could not explain how this happened. It was impossible to calculate the amount and wave length of light that would be radiated by a body from the amount and wave length of the radiation that it absorbed. Formulas were developed that worked for long wave lengths, and others that worked for short ones. But no single formula could be found that accounted for both. Something was wrong. A wave is a wave, no matter what its length, but here we had different wave lengths behaving as though they had no relation to one another, as though they represented entirely different types of natural phenomena.

Wave theory also failed to account for the photoelectric effect. It could not explain how some materials release electrons when you shine light on them. After the English physicist J. J. Thomson showed that electricity was nothing more than a moving stream of electrons, scientists looked closely at photoelectrons—the electrons in the current from a light-sensitive cell. What they saw did not make sense.

The energy in a light wave—as was seen through careful observation—knocks the electrons loose from the material in the light-sensitive cell. Light energy is, in other words, converted into electron energy in the photoelectric effect. According to classical mechanics, the brighter the light was, the greater would be the energy of each electron knocked loose. This was exactly what happened with ocean waves. Big ocean waves carried driftwood along faster than small ones did. A large wave of light should have done the same thing; however, it did not. No matter how bright or dim the light on the cell was, the electrons knocked loose always came out with the same average speed. Bright light merely loosened a greater number of electrons. Changing the color of the light changed the energy of the electrons. A dim yellow light, for example, always imparted more energy than a bright red light. Wave theory was at a loss to explain why.

These were a few of the discrepancies that plagued the scientists. When they started to poke into these corners, their complacency was shattered. Tidying up would not be enough. Those odd experiments could not be brought quite into line with standard theory; repeating them and refining them did not help. The theory was wrong. Minor discrepancies were discovered to be major, and they revealed flaws in the very foundations of science. The whole, beautiful edifice of nineteenth-century science collapsed.

The rebuilding of science began bright and early in the twentieth century. A shy, unkempt patent clerk in Bern, Switzerland, astonished the world with a theory that accounted well enough for the photoelectric effect, but stood classical theory on its head in the process. A few years later this same patent clerk, one Albert Einstein, would astonish the world even more with his theory of relativity, which outlines the best rules yet devised for understanding the operation of the universe.

Einstein's explanation of the photoelectric effect was

based on a concept first put forth in 1900 by Max Planck to explain the behavior of hot objects. Planck assumed that hot objects absorb energy in multiples of a minimum amount which he called "quanta." These quanta, according to Planck, could only be absorbed in specific numbers, the way a slot machine absorbs coins. It can be fed one quarter, one nickel, or any number of whole coins, but never half a quarter, half a nickel, or a quarter of a penny. This was a radical idea at the time.

Heat energy was supposed to be an electromagnetic wave, and waves are continuous. A wave cannot be divided into a minimum amount like so many marbles. Planck's equations, however, worked—they agreed with observation—and the wave theory of electromagnetic forces was seriously challenged.

When Einstein took up these ideas five years later to explain the photoelectric effect, he made the idea of quanta even more radical. His equations showed that light, heat, radio waves, all forms of electromagnetic radiation, must always be divided into these quantum packages, and he showed how this concept could explain the photoelectric effect.

According to Einstein, light had to be a stream of individual quantum packages. When this stream collides with a light-sensitive cell, each quantum package knocks a single electron loose. Bright light, the theory goes on to explain, contains more of these quantum packages and is able to free more electrons. The amount of energy in each package is, however, determined by the color of the light. This explains why different-colored light releases electrons at different energy levels.

Shades of old Isaac Newton and his discredited "infinitesimal corpuscules"! Here was a twentieth-century upstart telling the world that Newton was right all along, that light was made up of tiny corpuscles and not immaterial waves at all; and he had the mathematics to prove it.

But what of wave theory? Was all the work done under its influence to be thrown out the window? Not at all, said Einstein. Wave theory is much too accurate in describing many things about light to be abandoned. His answer to this paradox was simple, straightforward, and completely typical: Light, he said, sometimes acts like an advancing wave front and sometimes like a stream of tiny particles.

As can be imagined, the scientific world literally reeled with the impact of this wild notion. It was not too difficult to understand that light could be one or the other, particles or waves, but how in the world could it be both!

Among the bright young men who considered these strange notions was a French aristocrat named Louis de Broglie, Prince of Piedmont, a learned historian and an amateur physicist. He took to musing over Einstein's concept of light as tiny particles containing both mass and energy. He was also familiar with the evidence that light has wave motion as well, and from there he reasoned further: If light also has mass, which could make it some sort of aspect of matter, why should not all matter have wave motion? Why, indeed, should not matter consist entirely of waves?

De Broglie realized that he could never get anywhere with this idea unless he could somehow bring it down to earth with the aid of a mathematical model. He decided to concentrate on the electron, since this was one of the smallest particles known to man. If he could show that an electron could also sometimes behave like a wave, his problem would be solved.

Burrowing deep into the mathematical reserves of the time, de Broglie worked out a series of equations that turned out to be strange indeed. They showed something of the form and functioning of the electron's frequency. Since he was dealing with a wave frequency rather than a particle frequency, his electron emerged from its mathematical treatment, not as an orbiting particle, but as a kind of centralized throb.

His actual mathematical equations turned out to be interpretable in three ways: the electron can behave as (1) a concentrated rhythmic beat; (2) an explosive pulsation; or (3) both. De Broglie assumed the last: that an atom had not only a kind of localized core of stable matter, but also a cloud of electrons around it that broadcast an expanding pulsation "forever in step with it and extending all over the universe."

It is one thing to put forth a theory, no matter how convincing the mathematics, but quite another thing to prove it. When de Broglie published his theory in 1923, he realized that it would be difficult to prove. Neither he nor anyone else seemed to have any idea of how to conduct such an experiment. Yet, within a year and a half of publication, his findings were verified by experimental proof in what was almost a divine fluke.

A researcher named C. J. Davisson in the Bell Telephone Laboratories in New York was running through a routine experiment in April, 1925, that involved spraying a piece of nickel with a stream of electrons in a high vacuum. One day, everything seemed to go wrong. A flask of liquid air exploded near by, wrecking the apparatus Davisson was using and spoiling the nickel surface. He had to rebuild his entire apparatus and clean the nickel by heating it to a very high temperature before his experiment could be resumed. Not realizing that this treatment had changed the surface of the nickel, Davisson was amazed at the diffraction patterns made by the next stream of electrons—patterns which proved to have been produced by the very matter-waves predicted by de Broglie.

Davisson's discovery led to a Nobel Prize for both him and de Broglie. It not only verified de Broglie's theory of the wave nature of electrons, but even confirmed the extremely short wave length that he had predicted for the electron. This wave length is so much shorter than that of light that, after a little more than ten years, it provided man with an entirely new kind of optical tool which enlarged his vision of the microcosm five hundred-fold, through the marvelous new electron microscope.

These radical new ideas were also applied to the atom. By the end of the nineteenth century, the atom, too, was believed to have been thoroughly explored. The experiments of the English physicist J. J. Thomson and the American Robert A. Millikan had brilliantly revealed the innermost workings of the atom. They discovered the electron and identified it as the smallest charge of negative electricity. They weighed and measured it and found that it was a part of all matter. The atom, according to Thomson, consisted of a large, positively charged central core upon which the electrons were stuck like pins on a pincushion. Most of the weight and bulk of the atom was confined to this core.

It seemed a perfectly logical model of the atom. It agreed with observation and accounted for electric prenomena as observed in atoms at the time. In 1911, however, this model was shattered when another British physicist, Lord Ernest Rutherford of New Zealand, literally shot it down. He blasted the atom with particles given off by radioactive substances. When he aimed these particles at a sheet of gold foil, for example, he obtained amazing results. Most of his particles went right through the gold foil as though there were nothing there. Only rarely did one of Rutherford's alpha particles bounce off as though it had run into something solid. In fact, Rutherford was able to count no more than an average of one direct hit for every two million shots. There went Thomson's picture of an atom with a large central core. An atom consisted mostly of empty space—two million times more empty than solid.

From his findings, Rutherford developed an entirely new model of the atom. He pictured it as a miniature solar system. A tiny, heavy core corresponded to the central sun, and electrons whirled around this core like so many planets. The electrical attraction between the positive core and the negative electrons held the electrons in their orbits, just as gravity from the sun holds the planets in theirs.

This new model of the atom was hailed by scientists everywhere as the answer. Light, however, was to destroy this theory also. We know that every element gives off a specific color spectrum that is limited to characteristic color bands. According to Rutherford's model, an atom's spectrum should not behave like this at all. The forces of attraction exerted by the nucleus to hold the electrons in their orbits would also tend to slow them down. They should lose their energy gradually, releasing it as light of all colors; they should produce a continuous spectrum, not just specific bands. Worse still, it could be shown that, if the electrons lost energy, they would also have to move closer to the atomic core. They would spiral down into the nucleus, and the atom would disintegrate. Since our solar system has been in existence for at least six billion years, this theory was obviously wrong.

While Lord Rutherford struggled with the contradictions in his model of the atom, one of his students, a talented young Dane named Niels Bohr, decided to apply some of those wild ideas about particles and waves that he had heard so much about to these recalcitrant atoms. He seized on Einstein's concept of quantum—that light energy comes only in packages of a certain minimum size that cannot be divided.

In applying these ideas to the atom, Bohr assumed that electrons cannot lose energy gradually, but only in jumps, one quantum of energy at a time. This meant that each atom had certain specific orbits for its quota of electrons. An electron could jump from one permitted energy level or orbit to another, but could never be in-between. The specific energies released by these quantum jumps corresponded to the specific lines of color emitted by the atom. Since electrons were restricted to certain permitted orbits, they could never spiral into the nucleus.

With the Bohr model of the atom, the basic rebuilding of the collapsed edifice of science was completed. Bohr's model, of course, has since been refined and altered, but his basic ideas remain in current atomic theory. By 1926, this revolution in thought was all but won. The precocious young men who pulled off this coup went on to spend their adult years consolidating their victory. Conflicts in the mathematics had to be resolved; experiments had to be performed, repeated, and refined. Outside the physical societies, however, not many people knew about it. But the world of science would never be the same again.

Newton's classical concepts of mechanics were shown to be inadequate to explain the workings of the universe. In their place came a bewildering maze of abstruse mathematical formulas and strange ideas. Energy and matter,

*   In Wonderland   *   113

waves and particles, were shown to be aspects of the same thing. And they were governed by statistical laws—the mathematics of probability—that predicted what was most *likely* to happen, rather than what *would* happen.

The old laws of physics were not like this at all. If you threw a stone or fired a cannon, the old laws told you how high it would go and where it would land in numbers that were exact. There was no concern with probabilities, no ten-to-seven chance that it might land here or five-to-one chance that it could land there. Engineers built bridges confident that the girders would hold. Now, along came these bright young men who, in effect, said: Yes, the odds are that the metal will hold, but there is a chance—very, very slim—that all the atoms may move apart and let the bridge drop.

The laws of quantum mechanics, however, describe the actions of the physical world more accurately than the old laws did. The new theories predict what happens to atoms and parts of atoms when we do things with them. They show what happens when you shine light on a sensitive cell. They showed scientists how uranium atoms can be split, for example, and, they ushered in the atomic age.

Quantum mechanics corresponds with the physical world. It explains the actions of marbles, ocean waves, electrons, light quanta, everything—and it does so better than was ever done before. When applied to material things, the wave part of the law drops out, becoming so small as to be negligible. When applied to waves, the material part drops out.

If this is all very bewildering to you and outrages your common-sense view of things, the scientists merely answer that this is the way it works. Learn the mathematics, they say, and you will see how beautifully clear it all is: if you do not learn the mathematics, you will just have to take our word for it.

Energy and matter are the same thing in a universe ruled by statistical laws, where we can never say for certain where or what this or that is; this is our legacy from modern science. In all of this uncertainty, there is one value against which everything is measured, one point in relation to which everything else is relative—*the speed of light.*

11

THE UNIVERSAL CONSTANT

$$E = mc^2$$

Since the first atom bomb was exploded in Alamogordo, New Mexico, in 1945, this formula has become familiar to almost everyone. It is a mathematical expression of the relationship between energy and mass. It is basic to modern atomic theory and physics and explains the process by which the sun generates the enormous stores of energy that it lavishes on the earth. What this formula says is that energy and mass are two aspects of the same thing and that the energy potential of any particular mass can be computed by multiplying the mass by the speed of light squared. Because this speed of light is so great, we can see how much energy is bound up in mass.

So, here we have light once more, right in the middle of the relativity theory. This is not surprising, because certain discrepancies in light theory led directly to the formulation of Einstein's relativity concepts. Let us see how this came about.

As we have already seen, by the end of the nineteenth century the wave theory of light was accepted by most scientific authorities. The equations formulated by James Clerk Maxwell accurately described the observed behavior of light, granting one assumption: that light was propagated in a medium called the ether. Just as waves in the ocean are propagated in water and sound is propagated in the atmosphere, so, the scientists theorized, electromagnetic waves are propagated in the ether. Of course, no one had ever seen or felt this ether, but this fact alone did not matter. All we had to do, according to the scientists, was to develop an apparatus sensitive enough to detect this admittedly flimsy something.

The ether theory was also useful because it provided a fixed frame of reference against which motion could be measured. This was a good thing, too, because motion, which at first glance seems simple enough, is actually a difficult thing to pin down. Our bodies, for example, have no way of sensing motion. We feel changes of motion or changes in direction, but motion itself is not discernible. You can prove this to yourself with a little thought. We know that the earth is spinning on its axis at a rate of about 1,000 miles per hour; it is moving around the sun at a rate of about twenty miles per second; the sun is moving around the center of the Milky Way galaxy at a rate of about 600 miles per second; our galaxy is moving away from distant galaxies at a rate that approaches the speed of light; and neither you nor I can feel a single bit of this motion. Indeed, it was not until after the most painstaking kind of deductions that scientists finally acquired the knowledge of this complicated motion of the earth.

The Universal Constant

How then, asked the physicists and the mathematicians, do we measure motion? One method, with which we are all familiar, is to measure it against a fixed frame of reference. For example, you can calculate the speed of motion of a car in reference to how far it moves across the earth in a given period of time. In this case, the earth is the frame of reference against which we can measure the motion of the car. Scientists realized, however, that this kind of measurement was actually quite superficial.

When two trains are standing side by side in a railroad station, it is not unusual for a passenger in one of them to suddenly feel that he is moving backward when he sees the train beside him start to move ahead. The fact that he has the station as a frame of reference enables him to decide that his train is actually standing still while the other is moving. Now, imagine a condition where there is no convenient frame of reference against which to judge your movement. Say you are flying high above the clouds in a balloon that passes another balloon. It would be impossible for you to be certain when the two balloons pass each other, whether one or both are in motion, or to decide on the direction of the motion. Both balloons might be moving in the same direction relative to the earth below, except that one is moving faster than the other. When it overtakes the other, you could not say for sure whether you overtook the other balloon or whether it was moving in the opposite direction and merely passed you by.

When we begin to consider the motion of the stars, it becomes even more difficult to decide which is at rest and which is in motion. There are no stations or landmarks in the empty gulfs of space against which the astronomer can measure motion. He could consider the speed of the sun in relation to the fixed stars, but are there any stars that are really "fixed," or is there anything in the universe that we can use as a frame of reference?

Velocity could be shown to be absolute only if we could find one frame of reference for which the laws of nature would appear simpler to observers at rest within this particular frame than for those observers using any other background. If such a frame should exist, it could be singled out by any observer anywhere in the universe, and motion referred to it would be absolute. Newtonian mechanics could furnish no such convenient reference. With Maxwell's equations, however, scientists believed that they had discovered this universal reference in the ether in which electromagnetic waves were propagated.

There was only one drawback: the concept of the ether was only a theory with no observational proof to confirm it. Scientists all over the world attempted to prove its existence, without much success. Finally, in 1887, two American physicists named Albert Michelson and Edward Morley developed a precise experiment that should have demonstrated the drift of the ether.

To measure this ether current, Michelson and Morley devised an apparatus by which two light beams, reflected through a series of mirrors, were to cross the ether drift in opposite directions. According to light-wave theory, the light ray moving with the drift should arrive back at the source sooner than the ray that moved against the ether.

It should have, but it did not! Both rays arrived back at the source at exactly the same time. The experiment was repeated again and again to take account of all possible errors. Refinements were added so that a deviation as small as one-fifth kilometer per second would have been detected. You can appreciate the accuracy of this apparatus when you realize that light moves at a speed of 186,282 miles per second. One-fifth of a kilometer per second, then, represents an incredibly tiny deviation.

Despite the accuracy of the apparatus, the results were still negative. To all appearances, the earth stood permanently at rest within a stationary ether. This was, of course, impossible. We all know that the earth speeds around the sun at a rate of nearly twenty miles per second. It would be too much of a coincidence to expect the ether to move with the earth at exactly the same rate of speed.

Scientists were, however, reluctant to give up their concept of ether which provided such a convenient frame of reference for measuring motion. Attempts were made to explain away the findings of the Michelson-Morley experiment through mathematical reasoning. The Irish mathematician George Francis Fitzgerald and the Dutch physicist Hendrick Antoon Lorentz arrived independently at an identical solution to this problem.

Both men came up with the fantastic notion that lengths must contract in the direction of motion by an amount that accounted exactly for the fact that both light beams arrived back at the source at the same time. If we think of the rival rays as having a normal speed of five units per second (with each unit equal to about 37,200 miles per second) in an ether current of four units per second, then a contraction of length to 3.5 their original size as the rays move in the direction of the ether would explain why they returned to the same spot simultaneously. By projecting this rate of contraction at increasing speeds, the Lorentz-Fitzgerald equations show that a point must be reached where contraction becomes infinite. This point is reached at the speed of light.

Another experiment, performed by the English physicists Joseph John Thomson and W. Kaufman, showed that electrons moving at a very high rate of speed apparently increased in size in proportion with their increase in speed. Again, when this rate of expansion was projected, it was found that the electron would reach a limit to its expansion where it would have infinite size. This point is also reached at the speed of light.

Although the Lorentz-Fitzgerald equations explained away the contradictions in the Michelson-Morley experiment, other empirical facts, including the expanding electrons of Thomson and Kaufman, indicated that the difficulties revealed were not isolated failures of classical

ALBERT EINSTEIN (1879-1955)

The Universal Constant

physical theories, but actually pointed to serious flaws at the basis of classical theory.

It was left to Albert Einstein to proclaim, in his revolutionary paper of 1907, that the difficulties involved in all the classic experiments with light and electrons could best be solved by scrapping the traditional concepts of space and time. Einstein based his theory on two new axioms that were far different from Newton's axioms of motion, on which classical mechanics was based. These new axioms as described by Einstein are:

> 1. The principle of relativity, which states that the laws of physics are the same whether stated in one frame of reference or in any other moving with uniform velocity relative to the first.
> 2. The principle of *Celeritas*, which holds that the speed of light is constant. Its value remains the same when measured by all observers throughout the universe, and the speed is independent of its source.

The first postulate says, in effect, that it is impossible to define absolute motion, because to do so we must single out a special frame of reference—like the ether—and show that physical laws pertain to this frame only and to no other. We would have to show, for example, that a ball thrown in a resting train would react differently from a ball thrown in a moving one in respect to an observer in that train.

As for the second axiom, if we consider the speed of light a physical law, its constancy would follow from the first postulate: the laws of physics are the same in one frame of reference or in any other moving with uniform velocity relative to the first.

The constancy of light can also be demonstrated experimentally. We know that a light wave is governed by the same laws that govern all other waves. Thus, we can show that the speeds of sound waves are independent of their source. A supersonic plane, for example, is seen before it is heard. The velocity of the plane does not add one iota to the speed of the sound it produces. The series of sound vibrations travels out into the air with a certain characteristic speed whether the source is moving or at rest.

The same principle applies to light waves. Their speed is independent of their source. A ray of light coming from a speeding jet plane, for example, moves no faster than a ray of light coming from a stationary lamp. Light, no matter what its source or where in the universe it may be, always moves at the same speed.

On the basis of these two axioms, Einstein formulated both his special and general theories of relativity. The consequences of his revolutionary concepts are familiar to all of us. They have enabled scientists to unleash the energy tied up in the atom, and they have shown the way to the construction of complex transistors and tunnel diodes. As we have seen, the speed of light is basic to Einstein's theory; it is the main foundation on which the edifice of relativity has been constructed. Let us see now how scientists measure this all-important velocity.

In Chapter 1, we saw how Galileo tried to measure this speed by flashing lantern lights across distant hilltops. His method, of course, proved far too clumsy to actually measure a value as high as the speed of light. Later, the Danish astronomer Olaus Roemer was able to make a more accurate estimate by timing the eclipses of the moons of Jupiter. Though his method was more accurate than Galileo's, it still provided only an approximation because the astronomical distances were not known with sufficient precision to really refine the figures Roemer computed.

Scientists are, however, a stubborn breed of men. They seldom abandon a problem until it has been solved. Over the years, scientists devised ingenious methods for timing the speed of light, until today this value is known within an accuracy limit of a tenth of a mile per second.

The first relatively accurate mechanical measurement of light was accomplished in the 1840's by a French scientist named Armand Hyppolyte Louis Fizeau. He devised an apparatus that measured the speed by reflecting light back and forth between the teeth of a rotating wheel. Fizeau's method was improved by another Frenchman named Léon Foucault, who substituted a revolving mirror for the toothed wheel.

Foucault's experiment showed quite definitely that light moves faster in air, thereby effectively ruling in favor of the wave theory. But, as far as the speed of light was concerned, Foucault's apparatus proved remarkably accurate. Only technical refinements were needed to narrow it to its present established value of 186,282 miles per second, which has been repeatedly checked to within a fraction of a mile per second. The latest method, suggested by a Russian physicist in 1958, requires nothing more than an electronic flash tube, mirror, and photoelectric cell. In this method, the mirror serves only to reflect the flash from the tube back to the photocell, which is rigged to activate the flash over and over at a frequency that, by simple calculation, gives the speed of light.

More recently, in a triumph of technical perfection, an actual ray of light has been photographed in midpassage. This feat was performed by Dr. A. M. Zarem of the Stanford Research Institute, using a pair of electrically controlled, polarized filters coupled with a similar camera-shutter system that could expose its film for only one hundred-millionth of a second. With this apparatus, Dr. Zarem was able to capture a beam of light about ten feet long, freezing its presence on the photographic plate like a veritable fleeting glimpse of time itself, while both ends of the beam could be seen fading away into the surrounding darkness.

Here, then, is a beam of light finally impaled on a photographic plate. Somehow, it does not look impressive, rather like a stuffed lion when compared to a living one, roaming proudly through some veldt. Light, as we have seen, is more than just a beam of brightness. Indeed, it is fundamental to all matter, to all energy, and to all the interactions that give palpable shape to our universe. It has enabled astronomers to peer into the very heart of a star glowing billions of light years away. And it has given the physicist a wedge with which to pry open the secret sanctum of the atom.

Light is the marvelous glimmer of knowledge, and the Biblical ancients who gave its creation priority with the heavens and the earth were not far from wrong in their intuitive realization of the importance of light to the life of man.

GLOSSARY

ABERRATION OF LIGHT. A small apparent change in the position of a heavenly body caused by the motion of the earth. The effect is like that of a speeding car moving through a vertically falling rain. To the driver, the rain appears to slant almost horizontally against his windshield, and to the astronomer on the rapidly moving earth, the light from the stars appears to be coming at an angle. To overcome this aberration, the astronomer must aim slightly ahead of the true position of the star, just as a hunter "leads" a flying duck with his gun.

ABSORPTION BANDS. Dark bands on distended spectrum of the sun caused by loss of power in particular frequencies when, for example, sodium light passes through a cloud of vaporized sodium. The light is absorbed in the process.

ACHROMATIC. Colorless. An achromatic lens refracts light without breaking it up into its component colors.

ADVANCING WAVE FRONT. An imaginary surface made up of all the points reached at any given moment by a wave or vibration in its advance.

ANGSTROM UNIT. A measure consisting of one one-hundred-millionth of a centimeter. This is the unit used in measuring the length of light waves. Named after A. J. Ångström (1814-1874), Swedish physicist.

ASTIGMATISM. Structural defect in the eye lens that prevents light rays from meeting in a single focal point so that indistinct images are formed.

BETA RAYS. Rays given off by radioactive atoms, consisting of electrons that move with velocities that vary from 30,000 to 180,000 miles per second.

C, for *Celeritas*, the speed of light.

CHROMATIC ABERRATION. The appearance of a ring of color around the lenses of a telescope. It is due to the property of a lens' causing the various colors in a beam of light to be focused at different points, thus causing the spectrum to appear.

CHROMOSPHERE. Reddish layer of incandescent gases around the sun that is visible during an eclipse.

COHERENT LIGHT. A beam of light whose wave frequencies are all the same and parallel.

COLOR. The property of light waves of a particular frequency and length. The primary colors of the spectrum are red, orange, yellow, green, blue, indigo, and violet.

DIFFRACTION. The breaking-up of a ray of light into dark and light bands, or the colors of the spectrum, caused by the interference of one part of a beam by another when the beam is deflected at the edge of an opaque object or passes through a narrow slit.

DUTCH TRUNKS. Common name given the first binoculars made by Hans Lippershey in Holland.

ELECTROMAGNETIC SPECTRUM. The entire range of radiation that extends from cosmic rays whose wave lengths are on the order of .000,000,000,001 cm. to radio waves broadcast by alternating current, whose wave lengths are on the order of 1,000,000 cm.

ETHER. A hypothetical, invisible substance that fills all space, serving as a medium for the propagation of light waves and other forms of radiant energy. Its existence has been disproved.

※ *Glossary* ※

FLUORESCENT. A substance that emits light only while it is being excited by electromagnetic radiation. A fluorescent light is made up of a tube coated with fluorescent powder. Some mercury is added, and then the tube is filled with an inert gas. When an electric spark is passed through the tube, it causes the mercury vapor to give off ultraviolet radiation. This radiation, in turn, causes the fluorescent coating to glow.

FRAUNHOFER LINES. Absorption bands in the spectrum of the sun. Named after Joseph von Fraunhofer (1787-1826), who first observed and plotted them accurately.

GAMMA RAYS. Rays emitted by radioactive substances having a shorter wave length than X Rays.

HALE TELESCOPE. The 200-inch reflecting telescope at the Mount Palomar Observatory. The largest telescope in the world.

HALO. A ring of light that seems to encircle the sun, moon, or other luminous body caused by the diffraction of light through vapor.

HELIOGRAPH. A device for taking photographs of the sun.

HELIOSCOPE. Either a telescope or a device fitted to a telescope enabling one to look at the sun without hurting his eyes.

HIGH FREQUENCY. Radiation with a relatively high oscillation period, generally higher than 20,000 cycles per second.

HOLOPHOTAL. Having the quality of reflecting or refracting all or most of the light from a source.

IMAGE. A visual impression of something produced by reflection from a mirror or by refraction through a lens.

INCANDESCENT. Glowing with intense heat, as a stove that becomes red hot or even white hot.

INFRARED RAYS. That portion of the electromagnetic spectrum that falls just beyond the visible portion in the red end. These rays are longer than light waves, but shorter than radio waves.

Glossary

Ion. An atom or group of atoms which becomes charged with electricity when an electrically neutral atom or group of atoms either loses or gains an election.

Ionosphere. The outer part of earth's atmosphere, extending beyond the stratosphere, consisting of constantly changing layers of ionized molecules and atoms.

Lens. A piece of glass or other transparent substance shaped so that it bends light rays as they pass through.

Libration. An actual or apparent irregularity in the motion of a stellar body, such as the moon, planets, or stars.

Light. That portion of the electromagnetic spectrum which we can perceive with our eyes, falling in a range between 4,000 and 7,000 angstroms.

Light Year. Unit of astronomical distance equal to the distance that light travels in a year, about 6,000,000,000,000 miles.

Luminescence. The emission of light through the absorption of radiant energy, rather than through incandescence. Any cold light.

Luminous Flux. The rate of flow of light radiation.

Microscope. An instrument consisting of a system of lenses for making very small objects appear larger so that they can be seen and studied.

Mirage. An optical illusion caused by the refraction of light through layers of air of differing temperatures and densities by which a ship, an oasis in the desert, etc. appear to be very near and, often, upside down.

Mirror. A smooth surface, generally glass or metal, which reflects light.

Nebulae. A system of a large group of stars that revolves around a common center of gravity.

Novae. Exploding stars.

Objective Lens. The lens in either a microscope or a telescope that is closest to the object being observed.

Glossary

OPTICS. The branch of physics dealing with the nature and properties of light.
OSCILLATION. The variation between minimum and maximum values, as in an alternating electric current.
PARSEC. A unit of measurement for astronomical distances equal to 3.26 light years.
PHOTOELECTRIC EFFECT. The production of electricity by the action of light upon certain light-sensitive substances.
PHOTOMETER. An instrument used for measuring the intensity of light.
PHOTON. A quantum of light energy. Light, according to Einsteinian theory, is a stream of photons.
PHOTOSPHERE. The white-hot envelope of gas that surrounds the sun.
POLARIZED LIGHT. Light whose individual rays all move in the same direction.
QUANTUM. The elemental unit of energy. Quantum theory holds that energy is not absorbed or radiated continuously, but in definite, discontinuous units of quanta. The photon is the quantum of light.
RADIATION. The process by which energy in the form of heat and light is emitted by atoms and molecules as they undergo internal change.
RADIOACTIVITY. The emission of radiant energy in the forms of alpha, beta, and gamma rays and of light by the disintegration of an atomic nucleus.
RADIO WAVE. An electromagnetic wave having a frequency of more than 15,000 cycles with wave lengths ranging from one meter to a thousand kilometers.
REAL IMAGE. An image created by the actual meeting of light rays at a point.
REFLECTING TELESCOPE. A telescope with a concave mirror at the lower end of a tube which receives the light from an object and reflects it to a focus near the top of the tube.

Glossary

REFLECTION. The throwing back by a surface of sound, heat, or light.

REFRACTING TELESCOPE. A telescope in which a system of lenses causes light waves to focus, forming a magnified image of the object being viewed.

REFRACTION. The bending of a ray of light as it passes through a lens.

RELATIVITY. The theory of the relative, rather than absolute, character of motion, velocity, mass and of the interdependence of matter, time, and space, as developed and mathematically formulated by Albert Einstein.

SCHMIDT TELESCOPE. A type of reflecting telescope that can keep the edges of the image in sharp focus.

SCINTILLATION. The flash of light made by a ray or a particle from a radioactive material striking a crystal detector.

SPECTROSCOPE. An optical instrument used for forming spectra of light so that they can be measured and studied.

SPECTRUM. The series of colored bands diffracted and arranged in the order of their respective wave lengths by passing a white light through a prism or other diffracting medium. The spectrum shades continuously from red, produced by the longest visible waves, to violet, produced by the shortest visible waves.

SPECTRUM ANALYSIS. A method for determining the chemical make-up of a substance by heating it to incandescence and then passing this light through a spectroscope. All elements show characteristic bands in the spectrum. Gold, for example, "broadcasts" at 2427.95 angstroms, whereas only sodium vibrates at 5889.95 angstroms.

SUN. The incandescent body of gases about which the earth and all the other planets revolve and which furnishes light, heat, and energy for the solar system. It is the star that is nearest to the earth, with an average distance of about 93,000,000 miles. The sun has a diameter of approximately

865,000 miles, a mass about 322,000 times greater than Earth, and a volume about 1,300,000 times greater than that of Earth's.

SUN SPECTRUM. The series of colored bands produced when sunlight is passed through a prism.

TELESCOPE. An optical instrument that magnifies distant objects, used in astronomy to study the stellar bodies.

TEMPERATURE. A measure of the average molecular motion of a substance, generally relating to the hotness or coldness of the thing being measured.

ULTRAVIOLET LIGHT. That portion of the electromagnetic spectrum that lies just beyond the violet end of the spectrum. These rays of shorter wave length than visible light are important to our well-being. They react with ergosterol in the skin to produce Vitamin D.

URANIUM. A very hard, heavy, moderately malleable, radioactive metallic element.

VACUUM. A space with nothing in it.

WAVE. Any of a series of advancing impulses set up by emission from a molecule or atom that either gains or loses electrons.

X RAY. A nonluminous electromagnetic radiation of extremely short wave length, generally less than 2 angstroms.

ZODIACAL LIGHT. A faint, elliptical disk of light around the sun, usually seen in the west during or after twilight and in the east before daybreak.

CAPSULE BIOGRAPHIES

ÅNGSTRÖM, ANDERS JONAS (1814-1874), Swedish physicist who did pioneer work on the measurement of light waves. The angstrom unit (a hundred-millionth of a centimeter), named after this scientist, is the accepted unit for measuring wave lengths of the electromagnetic spectrum. Visible light falls in a range between 4,000 and 7,000 angstroms, though there is some variation in individual eyes. Some people can see slightly different ranges.

ARISTOTLE (c. 384-322 B.C.), Greek philosopher whose studies of natural phenomena provided the foundation of scientific learning throughout the Middle Ages. Aristotle was one of the first philosophers to give serious thought to light. He theorized that white light was pure and transcendental, whereas color was merely an infection of pure white light with earthly properties. Aristotle also speculated on the speed of light, believing that light moved at a finite, measurable speed.

BACON, ROGER (c. 1214-1294). Known as the "Admirable Doctor," this English philosopher and scientist was a member of the Franciscan Order at Oxford. His scientific researches, however, led to accusations of heresy. He was exiled to Paris for ten years, and at the close of his life was imprisoned for another fourteen years by church authorities. Bacon was one of the founders of the modern scientific movement, whose motto, "Experiment! Experiment!" was a rallying point for those who sought to learn the truth about nature. Among Bacon's accomplishments were the preparation of a more accurate calendar; experiments in chemistry, including the preparation of gun powder; and the encyclopedic *Opus Majus*, a treatise that attempted to catalogue all of man's knowledge. He also did important work in optics, inventing the magnifying glass and developing new views on refraction.

BOHR, NIELS (1885-1962). This Danish physicist was one of the outstanding scientific figures of our century. Born in Copenhagen and educated at Cambridge in England, he worked with Ernest Rutherford and Sommerfield in the development of quantum mechanics as applied to atomic phenomena. Escaping German-occupied Denmark during World War II in an exciting cloak-and-dagger operation, he came to America, where he worked on the atomic bomb. Bohr's principal contribution to atomic theory lay in his concept of the atomic structure. By applying quantum theory—developed by Albert Einstein and Max Planck in their studies of light phenomena—to the atom, Bohr was able to explain the spectrum of the various atoms and thereby come to a more accurate knowledge of atomic mechanics.

BRAHE, TYCHO (1546-1601), Royal Astronomer of Denmark who established a famed observatory on the island of Hven in the Baltic Sea. Brahe constructed the most accurate measuring and sighting instruments of his day. Although he rejected the Copernican system and held that the planets and the sun revolved around the earth, the accuracy and scope of his measurements of stellar movements provided a foundation on which later astronomers built. Brahe was so awed by the heavens that it was said that he always wore his best robes when he studied the stars, in deference to their majesty.

DESCARTES, RENÉ (1596-1650). Descartes was noted equally as

* *Capsule Biographies* * 135

a mathematician and a philosopher. Educated as a Jesuit and trained for an army career, Descartes developed his mathematical gifts early and became, not only the founder of modern philosophy, but also the founder of analytic geometry, demonstrating how lines and curves can be expressed as algebraic equations. He did pioneer work in the study of light and proved that the rainbow is a result of refraction of sunlight through tiny drops of water.

EINSTEIN, ALBERT (1879-1955). This Jewish-Austrian-Swiss-American theoretical physicist was one of the giants of science. His ideas created the greatest revolution in scientific theory since Copernicus proved that the sun, not the earth, was the center of our solar system. Einstein's "special" theory of relativity was developed while he worked as an examiner for the patent office in Switzerland; his "general" theory, involving an entirely new concept of gravity and the relationship between matter and energy, was completed about 1915. These two theories provide the best rules yet devised by man for understanding the working of the universe. These two theories led to the atomic bomb, on the one hand, and to an explanation of the perturbations of the orbit of the planet Mercury, on the other. Among other things, Einstein explained the principles of the photoelectric effect, the influence of gravity on light, and the Brownian movement. A humanitarian of broad interests, Albert Einstein was a champion of liberal causes throughout the world.

FRAUNHOFER, JOSEPH VON (1787-1826). A German optician, he was the maker of the finest lenses and telescopes of his time. He gave the name "Fraunhofer lines" to the dark lines of the solar spectrum. Although he was unable to explain these lines, he plotted more than five hundred of them in a distended spectrum of the sun. Fraunhofer made important improvements in telescopes, prisms, and other optical instruments and invented the heliometer, a micrometer, and a diffraction grating for the measurement of the wave lengths of light rays.

GALILEI, GALILEO (1564-1642). One of the founders of modern science, Galileo's triumphs and defeats are probably more generally known than those of most scientists. Galileo was the virtual founder of dynamics, anticipating Newton's famous three laws of motion, and was one of the first "modern" astronomers in that he used a telescope to study the heavens. He discovered, among other things, that the moon's light is reflected from the sun, that the Milky Way is made up of innumerable stars, and that the moon has monthly and annual librations. He was also one of the first scientists to attempt to measure the speed of light.

HERSCHEL, WILLIAM (1738-1822). A German astronomer who lived and worked in England, Herschel is probably best known for having discovered the planet Uranus. Herschel also experimented with light, discovering infrared rays when he placed a thermometer just outside the red end of the spectrum. When the temperature rose, Herschel theorized that this was caused by unseen rays.

HERTZ, HEINRICH RUDOLF (1857-1894). This German physicist was responsible for the principal developments which led to the invention of radio. Following the theories of James Clerk Maxwell, Hertz demonstrated the existence of radio waves. He measured their wave length and frequency and showed that they could be reflected, refracted, or polarizd in exactly the same way that light can be. Today, the behavior he described is known to apply, not only to light and radio waves, but to all the rays in the electromagnetic spectrum.

HOOKE, ROBERT (1635-1703). An irascible Englishman who quarreled with almost all the scientific figures of his day, Hooke was secretary of the Royal Society during a particularly fruitful period in the development of science. Hooke did important work in the development of both

* *Capsule Biographies* * 137

microscope and telescope and, together with Christiaan Huygens, developed the wave theory of light.

HUYGENS, CHRISTIAAN (1629-1693). Huygens was a gifted Dutch mathematician who, like Newton and Galileo, worked in the related fields of optics, astronomy, and mechanics. His *Treatise on Light*, published in 1678, proposed his famous wave theory of light, which he developed with the English scientist, Robert Hooke. Huygens also constructed telescopes, discovered one of the moons of Saturn, and proved that the rings of Saturn completely surround the planet.

JANSKY, KARL G. (1905-1950). An American physicist born in Norman, Oklahoma, who is best known for discovering the radio waves broadcast by distant stars, galaxies, and hydrogen clouds. This discovery laid the basis for the newly developed field of radioastronomy, which provides a valuable tool for the study of the heavens. Although Jansky made his initial discoveries in 1935 while working at the Bell Telephone Laboratories, his findings could not be utilized. The radio-receiving techniques of that time were inadequate to the study of this phenomenon. The development of radar in the course of World War II, however, produced radio-receiving equipment of a sensitivity and accuracy that made the study of these waves possible. As soon as the new instruments were trained on distant radio sources, it was quickly realized that Jansky's discovery had opened an entirely new avenue for the exploration of space.

KIRCHHOFF, GUSTAV ROBERT (1824-1887). This German physicist is credited with the establishment of spectrum analysis. He discovered the fact that, when a beam of light from a given element passes through a cloud of light from the same element, it is absorbed in the process, leaving only a dark line where it should have appeared on the spectrum. By applying his findings to the spectrum of the sun, he explained the existence of the Fraunhofer lines as bands of absorption, thus providing a vital clue to analyzing the chemical make-up of the sun.

LEEUWENHOEK, ANTON VAN (1632-1723). "The original microbe hunter" is a title often applied to this Dutch scientist. He is credited with the invention of the microscope and was the first man to see and describe the tiny living creatures which we call bacteria and germs.

MAXWELL, JAMES CLERK (1831-1879). Maxwell was a brilliant Scottish mathematician whose papers were read before the Royal Society of Edinburgh when he was only fourteen years old. He discovered that light waves were part of the electromagnetic spectrum, a series of alternating electromagnetic fields flowing through space. His famous work entitled *Treatise on Electricity and Magnetism*, published in 1873, not only explained the known facts about the movement of light waves, but also predicted the existence of radio waves which had not yet been discovered. Unfortunately, Maxwell's work was largely ignored until Hertz produced experimental evidence of the existence of electromagnetic waves longer than light waves in 1888. Maxwell also did pioneer studies of color and color perception.

NEWTON, ISAAC (1642-1727). Newton's *Principia*, published in 1687, is one of the landmarks of science. In this work, the great English mathematician and physicist set forth his theory of universal gravitation, bringing the entire visible universe, the earth and the heavens, into one indivisible whole for the first time in history. Born less than a year after Galileo's death, Newton's work embraced the achievements of all the great thinkers who preceded him and brought science across the threshold into the modern era. Besides his monumental work on gravity and motion, Newton also made important advances in many other aspects of the natural sciences. He discovered calculus, an indispensable mathematical tool for the study of motion; invented the reflecting telescope; and developed a "corpuscular theory" of light that has been a source of scientific controversy right down to our own century. Isaac Newton stands as one of the true giants of science, his work and ideas providing one of the important foundation stones on which the edifice of scientific knowledge has been erected.

PLANCK, MAX (1858-1947). Nobel Prize winner for physics in 1918, Max Planck is known today for his association with the quantum theory, which he began formulating in 1901. A quantum, according to his theories, is the smallest amount of energy found in nature, and all energy changes can be expressed in units of quanta. Planck was led to his theory through the study of phosphorescent light, which, when observed through a spectroscope, does not flow in a continuous line, as physics had previously taught, but in separate "bundles," or "quanta." His theories were applied a few years later by Albert Einstein to the problem of the photoelectric effect and played an important part in Einstein's "photon theory" of light.

POINCARÉ, JULES HENRI (1854-1912). A French mathematician and physicist, Poincaré proved Maxwell's theory that a light wave is a series of alternating electromagnetic fields flowing through space. He proved that these fields change direction, or alternate, 1,000,000,000,000,000 times per second.

ROEMER, OLAUS (1644-1710). A Danish astronomer noted for his successful measurement of the speed of light. By measuring the apparent difference in the speed of eclipse of the moons of Jupiter as timed from opposite sides of Earth's orbit, Roemer calculated the speed of light. Although his figure was not completely accurate, it came remarkably close to our currently accepted time. Roemer was able to accomplish this calculation through the invention of a pendulum clock that enabled him to time heavenly phenomena more accurately than ever before

✳ Capsule Biographies ✳

ROENTGEN, WILHELM KONRAD VON (1845-1923). A German physicist, he is the discoverer of X rays, which proved to be high-frequency electromagnetic waves with a wave length far smaller than that of visible light.

SCHRÖDINGER, ERWIN (1887-1961). A German physicist who won the Nobel Prize in 1933. Schrödinger is best known for his mathematical model that could demonstrate both the particle and wave nature of light.

WOLLASTON, WILLIAM (1766-1828). An English physicist who first noticed the absorption bands on the spectrum of the sun. One of the discoverers of the spectroscopic language of the atom.

INDEX

aberration of light, 125
absorbtion bands, 78-79, 125
achromatic lens, 125
advancing wave front, 33, 125
Alhazen, 29-31
alpha rays, 66-67, 111
aluminum, on telescope mirrors, 56-57
ammeters, 74
Ångström, Anders Jonas, 64, 133
angstrom unit, 64, 124
Aristotle, 4-6, 30-31, 36, 133
astigmatism, 125-126
astronomy, 9, 45-50, 54-59, 82-91; *see also* stars; telescopes
atomic age, 113
atomic bomb, 96, 115
atomic energy, 122
atoms, Bohr model, 96, 112; collapse of, 18-19; Rutherford model, 111-112; Thomson model, 110-111; wave lengths, 97

Bacon, Roger, 4, 133-134
beam of light, photographed, 123-124

Bell Telephone Laboratories, 84-85, 94, 100, 109
bending, *see* refraction
beta rays, 67, 125
binoculars, 44
black light, 94
Bohr, Niels, 4, 96, 112, 134
Brahe, Tycho, 134
Briggs, Leonard, 43
Broglie, Louis de, 4, 108-110

cameras, 53
canals, Martian, 58-59
cancer, 67-68
Capella, 86
Cassiopeia, 86
cathode tube, 66
Celeritas, principle of, 121, 126
chemistry, and lasers, 101
chromatic aberration, 33, 47-48, 126
chromium atoms, 98
chromosphere, 126
coherent light, 92-101, 126
cold light, 71
color, 5, 27-35, 49, 60-62, 72-81, 126

Index

condensation, of rain, 14; theory, 22-23
copper, spectrum, 77
Corning Glass Works, 48
corpuscular theory, *see* particle theory
correcting plates (lenses), 57
cosmic rays, 68
cost, of light, per pound, 3
crab nebula, 83
"curve" ball, in baseball, 34-35
Cygnus, 49
Cygnus A, 83-84, 88

dark stars, 86
da Vinci, Leonardo, 43
Davisson, C. J., 109-110
de Broglie, Louis, 4, 108-110
Dee, John, 43
de Hulst, Van, 87
della Porta, Giambattista, 43
Descartes, René, 29, 31, 134-135
diffraction, 104, 126
Dirac, Paul, 4
dust clouds, interstellar, 87
"Dutch Trunks," 44, 126

earth (planet), 16-18; motion of, 116
Echo I, 28
Einstein, Albert, 2, 4, 20-21, 37, 95-96, 106-108, 116, 120-122, 135
electrical phenomena, and light, 69
electricity, 13-14; as electron stream, 105
electromagnetic radiation, 63-70, 73-76, 102-106; *see also* radio waves
electromagnetic spectrum, 25-38, 60-70, 126
electron microscope, 110
electrons, 96-98; in atoms, 111-112; and color, 106; electricity as stream of, 105; energy levels, 96-98, 112; expansion of with speed, 120; wave length, 110; wavelike behavior, 108-110
elements, spectra of, 77, 111; in sun, 80
Empedocles, 4-6, 14
energy, atomic, 122; and color of light, 106
energy-mass formula, 115
Epsilon Aurigae stars, 10
ergosterol, 65
ether, concept of, 29, 116, 118, 121, 126
eyeglasses, 42-43
eye operations, with laser beams, 100-101

Faraday, Michael, 69
Fermi, Enrico, 4
Fitzgerald, George Francis, 119
Fizeau, Armand Hyppolyte Louis, 123
fluorescence, 98, 126
focus, of lenses, 53-54
Foucault, Léon, 4, 123
frames of reference, 116-121; *see also* relativity theories
Fraunhofer, Joseph von, 78, 135-136
Fraunhofer lines, 76-80, 127

Index

Fresnel, Augustin, 36-37

galaxies, 9-11, 57, 59, 116; in collision, 83, 88
Galilei, Galileo, 4-6, 8, 44-46, 59, 84, 122, 136
gamma rays, 67-68, 127
gas lasers, 100
General Electric, 100
gold, spectrum, 77
Gregory, James, 47
Grimaldi, Francesco Maria, 4, 31-32

Hale telescope, 48-49, 56, 58, 60, 88-89, 127
halo, 127
heat, beams from lasers, 100; radiation, 62-63, 71-81, 104-105, 107
heliograph, 79, 127
helioscope, 127
Herschel, William, 136
Hertz, Rudolph, 4, 69-70, 136
Hess, V. F., 67
holophotal, 127
Hooke, Robert, 4, 31-33, 36-39, 136-137
Hulst, Van de, 87
Huygens, Christiaan, 4, 32-33, 36-37, 137
Huygens' principle, 33
hydroelectricity, 14
hydrogen, 22-23, 80, 87

image, 52-55, 127
Inca, 16
incandescence, 75, 127
infrared photography, 64

infrared rays, 62-65, 100, 127
intensity, of light, 51-54
Io, 6-8
ion, 128
ionosphere, 128
iron, spectrum, 77, 80

Jansky, Karl G., 84-86, 137
Japan, 16
Javan, Ali, 100
Jodrell Bank Radio Observatory, 89
Jupiter, 17; moons of, 6-8, 46, 122

Kaufman, W., 119-120
Kepler, Johannes, 46
Kirchhoff, Gustav Robert, 78, 137-138

lasers, 94-101
Leeuwenhoek, Anton van, 138
Leiden, 87
lenses, 39-47, 52-57, 128; achromatic, 125
Leonardo da Vinci, 43
libration, 128
light year, 11, 128
Lippershey, Hans, 43-45
Lorentz, Hendrick Antoon, 119
Lorentz-Fitzgerald contraction, 119-120
Lovell, Bernard, 89
luminescence, 128

Magellanic Cloud, 10
magnifying glasses, 42, 52-53; *see also* lenses

144 * Index *

Maiman, Theodore H., 94, 98-99
Marconi, Guglielmo, 70
Mars, 17; canals of, 58-59
Maxwell, James Clerk, 4, 69-70, 102-104, 116, 138
Medicean satellites, see Jupiter, moons of
Mercury, 17
Michelson, Albert, 118-120
Michelson-Morley experiment, 118-120
Micrographia, 32
microscopes, 39, 128
milk, irradiated, 65
Milky Way galaxy, 23, 46, 83-85, 87-88, 116
Millikan, Robert A., 110
miniaturization of equipment, 101
mirage, 41, 128
mirrors, in telescopes, 47-49, 55; see also telescopes, reflecting
Mithras, 16
moon (earth satellite), 46, 54, 101; laser beam shot to, 101
moonlight, 27
moons, of Jupiter, 6-8, 46, 122
Morley, Edward, 118-120
Moseley, Henry Gwyn, 4
motion, absolute, 121; concept of, 116-121
Mount Palomar Observatory, 48-49, 56-58, 88-89

natural fluorescence, 98
Navy, U. S., radiotelescope, 89-90
nebulae, collision of, 88

Neptune, 17
Newton, Isaac, 4, 33-39, 47-48, 55, 60, 76, 104, 107-108, 112, 118, 138-139
Newtonian mechanics, 112, 118
Nineveh lens, 39-40
Nobel Prize, 110
Nuclear fusion reaction, in sun, 21-24

Oersted, Hans Christian, 69
Oort, Jan, 87
Optics (Ptolemy), 40
orbits, of electrons in atoms, 96-98
Orion's belt, 46

Palomar telescope, see Hale telescope
parsec, 129
particle theory, 31, 35-37, 48, 104, 107-108; see also wave theory
phosphorescent materials, 96-97
photoelectric effect, 95, 105-107
"photographic" telescopes, 55, 58-59
photography, 57-58, 64-65, 123-124
photons, 2, 96, 129
photosphere, sun's surface, 19-21, 129
Planck, Max, 37, 94-95, 107, 139
planets, 46
Pluto, 16-17
Poincaré, Jules Henri, 69-70, 139
polonium, 67
Porta, Giambattista della, 43
potassium, spectrum, 78

Index 145

potency, of light, 3
pressure, exerted by light, 28-29
prisms, experiments with, 34, 60-62, 73, 76
Ptolemy of Alexandria, 40-43
pyramids, 16
quantum, 94, 129; theory of, 37, 95, 113-114

Ra, 16
radar, 64, 86, 93
radiation, 129; electromagnetic theory of, 69-70; relation to temperature, 73-81
radioastronomy, 84-91
radiotelescopes, 82-91
radio transmitters, 63-64
radio waves, 63-64, 69-70, 82-91, 93, 104, 129; see also electromagnetic radiation
radium, 67
rainbows, 26-28, 62
"real images," 53-54, 129
reflecting telescopes, see telescopes, reflecting
refracting telescopes, see telescopes, refracting
refraction, 40-41
relativity, principle of, 121; theories of, 2, 20-21, 106, 115-116, 120-122, 130
resistance, electrical, 74
resonance, 97-98
retina, 52, 101
rings of Saturn, 46
Roemer, Olaus, 6-9, 122, 139
Roentgen, Wilhelm Konrad von, 66, 140
Royal Society, 32, 34, 36

ruby lasers, 98-100
Rutherford, Ernest, 4, 67, 110-112
Ryle, Martin, 88

S Doradus, 10
Sagittarius, 84
satellites, man-made, 28
Saturn, 10, 17; rings of, 46
Schawlow, Arthur, 94, 97-99
Schmidt telescope, 57-58, 130
Schrödinger, Erwin, 140
Seneca, 42
silver, spectrum, 77
Sirius, 86
sodium, spectrum, 77-78
solar sailing ships, 29
solar system, 16-17
sound, speed of, 121-122
sources of light, 13-25
space hazards, 68
space ships, 29
Spain, 44
spectacles, 42-43
spectroscopes, 77-78, 130
spectrum, electromagnetic, 26-38, 60-70, 126 (see also radio waves); Fraunhofer lines, 76-80; stars, 76; and temperature, 71-81
speed of light, 2-9, 40-41, 114-124; as universal constant, 114-124
speed of sound, 121-122
stars, classes, 10-11; colors, 49, 73; dark, 86; radio waves, 82-91; size and distance, 76; spectra, 76; temperature, 73-76

Index

statistical laws, for universe, 113-114
Stonehenge, 16
sun, 2-3, 21-24, 130; age of, 23; chemical composition, 78-80; expansion of, 24; hydrogen, 80; nuclear-fusion reaction of, 21-24; size, 76; spectrum, 78-80, 131; symbols, 15
surgery, with laser beams, 100-101
synthetic rubies, in lasers, 98-99

telescopes, 39, 43-59; Aracaibo radiotelescope, 90; Hale, 48-49, 56, 58, 60, 88-89; Jodrell Bank radiotelescope, 89; "photographic," 55, 58-59; radiotelescopes, 82-91; reflecting, 47-49, 55-58, 129; refracting, 33, 46-47, 56-57, 130; Schmidt, 57-58; "visual," 54-55, 58-59; Yerkes, 56
temperature, and spectra, 71-81
thermoelement, 63, 74
Thomson, J. J., 105, 110-111, 119-120
Townes, Charles H., 94, 97-99
transistors, 96, 122
Treatise on Light, 33
tunnel diodes, 122

ultraviolet rays, 62-63, 65-66, 131
universal constant, light speed as, 114-124
universe, hydrogen in, 22-23; scope of, 9-10; statistical laws for, 113-114
uranium, 67, 113, 131

van Leeuwenhoek, Anton, 138
variable resistors, 74
Veil Nebulae, 49
velocity, 118-120; *see also* speed of light
Venice, 42, 46
Venus, 17
"visual" telescopes, 54-55, 58-59
Vitamin D, 65
von Fraunhofer, Joseph, 78, 135-136
votive candles, 1

water, and refraction, 40
wave fronts, 33
wave lengths, atoms, 97; electromagnetic, 63-64, 104
wave theory, 31-33, 36-38, 104-108, 123; *see also* particle theory
weight, of light, 3
West Virginia radiotelescope, 89-90
white dwarfs, 10-11, 24
Wollaston, William Hyde, 65, 140

X rays, 66-67, 131

Yerkes telescope, 56
Young, Thomas, 4

Zarem, A. M., 123-124
zodiacal light, 131